The Social Fact

The Social Fact

News and Knowledge in a Networked World

John P. Wihbey

The MIT Press
Cambridge, Massachusetts
London, England

This book was set in Stone Serif & Stone Sans by Jen Jackowitz. Printed and bound in the United States of America.

Library of Congress Cataloging-in-Publication Data

Names: Wihbey, John P., author.
Title: The social fact : news and knowledge in a networked world / John P. Wihbey.
Description: Cambridge, MA : The MIT Press, 2019. | Includes bibliographical references and index.
Identifiers: LCCN 2018028559 | ISBN 9780262039598 (hardcover : alk. paper)
Subjects: LCSH: Journalism--Social aspects--United States--History--21st century. | Social media--Social aspects--United States. | News audiences--United States--History--21st century. | Mass media--Technological innovations.
Classification: LCC PN4888.S6 W55 2109 | DDC 071/.3--dc23 LC record available at https://lccn.loc.gov/2018028559

10 9 8 7 6 5 4 3 2 1

For Frank Wihbey, 1944–2010:
Librarian, Scholar, Dad.

Contents

Acknowledgments

This book finds its deepest sources in a network of people who have provided wonderful support and inspiring ideas. My work at Northeastern University's School of Journalism and Harvard's Shorenstein Center on Media, Politics and Public Policy has crystallized most of the major ideas presented here. Although mistakes are my own, I am indebted to my many colleagues and students at these institutions.

My profound gratitude to Nick Lemann and Tom Patterson for their intellectual guidance. Jonathan Kaufman, Dan Kennedy, David Lazer, Richard Parker, and Brooke Foucault Welles provided important advice. I would like to thank my research collaborators on various projects, including Kenny Joseph, Mike Beaudet, Matt Nisbet, Alison Head, Nicco Mele, Pedro Cruz, Aleszu Bajak, Mark Coddington, Laurel Leff, Meg Heckman, Michelle Borkin, Bud Ward, and my sister, Lynn C. Sweet. My research assistants Bridget Peery and Amy Stubbs and Northeastern librarian Brooke Williams were essential in helping to prepare the manuscript. Northeastern students Felippe Rodrigues, Matt Tota, and Adam Wiles also provided important assistance. I would also like to thank Steven Braun and Dan Zedek for their wonderful design help. Gita Devi Manaktala at the MIT Press has provided terrific editorial guidance and encouragement throughout. I am grateful to Kathleen Caruso, Melinda Rankin, and Mary Reilly for their vital editing work.

At my home institution, Northeastern University, my thanks go to Meryl Alper, Jane Amidon, Hilary Poriss, Dan Cohen, Sarah J. Jackson, Jeff Howe, Carlene Hempel, Matt Carroll, Dan O'Brien, Dietmar Offenhuber, David Tamés, Alan Schroeder, Walter Robinson, Jim Ross, Dina Kraft, Chuck Fountain, Susan Conover, and Gladys McKie. From my days at the Shorenstein Center, I owe many debts to Nancy Palmer, Jeff Seglin, Matthew Baum,

Alex S. Jones, Marion Just, Leighton Kille, and the late Jonathan Moore. My thanks, as well, to David Weinberger, Jeff Hermes, Andy Sellars, and Josh Benton. For their support during my time teaching at Boston University, I would like to extend my gratitude to Christopher Daly, Michelle Johnson, and Bill McKeen. I especially want to thank Duncan Watts, Matthew Gentzkow, Meredith Broussard, Josh Stearns, Nick Diakopoulos, Ingmar Weber, Sudhir Venkatesh, and Albert-László Barabási for their time and insights. The three anonymous reviewers of my draft manuscript made tremendous suggestions, and I am in their debt.

Northeastern's College of Arts, Media and Design (CAMD) and the NULab for Texts, Maps, and Networks have been supportive of the research for this book. My thanks to the Stanton Foundation, Knight Foundation, Carnegie Corporation of New York, Barr Foundation, Heising-Simons Foundation, Facebook, and the Association of College and Research Libraries (ACRL) for supporting various projects and research endeavors.

I count myself extremely lucky to have had many wise teachers through earlier years, including Thom Ingraham, Sandy Phippen, Leo Pelleriti, Al McLean, Scott Sehon, Paul Franco, Eddie Glaude, Matthew Stuart, Bill Watterson, Jonathan Freedman, Margery Sabin, Isobel and Michael Armstrong, Paul Muldoon, John Elder, David Huddle, Andie Tucher, James Carey, Robin Reisig, and Sandy Padwe.

My mother, Karen Wihbey, has been so generous and encouraging at every step of the way. Without the support and patience of Carrie Wihbey, this book never would have been finished. To you, Carrie, and to Lola, Max, Emmylou, and Amelia, I also dedicate this book.

Introduction

We have the physical tools of communication as never before. The thoughts and aspirations congruous with them are not communicated, and hence are not common. Without such communication the public will remain shadowy and formless, seeking spasmodically for itself, but seizing and holding its shadow rather than its substance.

—John Dewey, *The Public and Its Problems* (1927)[1]

Our media crisis is more complex than we typically imagine. Growing problems with credibility, trust, polarization, fake news, and the business model underpinning news production are all very real. But beneath these problems are structural rifts we are only beginning to discern. Forces are pulling apart norms and expectations that provide the context for news as we have known it. News is a form of culture, and the culture is changing. Technology fuels these changes, but they represent a fundamental shift in societal needs and values.

Overall, the public believes journalism remains crucial for democracy, although there is a broad sense that news media are performing this role poorly.[2] The public vaguely senses new cultural forces at work, but we have only inchoate language and categories through which to respond and understand. Hence a certain plaintive question now haunts us: What is going on with the news? Concerned citizens keep asking this question. Without adequate language, there is confusion and anger.

It is a sign of the times that we are now at a loss to define news. News may include a mobile video of a dying protestor or a nightly TV segment about a monster storm swamping a city. It may be a remixed, overdubbed, satirical video of a politician speaking or a quip on Twitter from an activist—or even a sitting president. The definition of news floats around unmoored.

Correspondingly, the public's evaluations of what news does, its quality and its effectiveness, floats around. It no longer makes any sense to speak of *the media*, although we often fall back on such language.

This book is about the structural crisis destabilizing news and how civic-minded media producers should respond. It explores how the structure of news, information, and knowledge, and their flow through society, are changing, and it projects forward ways in which news media can demonstrate the highest possible societal value in the context of these changes.

The overarching discussion here is about news, but this book focuses a great deal on the deeper factors that will shape media and the context of production in the future. To understand these factors, we must delve into both network science and the interplay between information and communication technologies (ICTs) and the structure of knowledge in society. These are complicated matters—and we can draw only tentative conclusions—but such exploration is necessary for us to grasp what really may be going on with news, and where current trends may lead us. This book's insights, in any case, are meant to be accessible to all those concerned about the future of news and public affairs.

The reader will, I hope, learn a substantial amount about the underlying patterns that characterize our increasingly networked world of information, with its viral phenomena and whiplash-inducing trends, its extremes and surprises. I aim to put these patterns in their proper context through journeys into media history and into the broader story of our current information revolution. By looking more carefully at the engineering behind information technologies, from the telegraph to the likes of Facebook and Google, this book tries to situate questions about news media practice in broader perspective. The shaping architectures of communications technologies and the systems and ecosystems they create matter a great deal for news.

It has remained unclear how the traditional media world can be reconciled with the world of social, peer-to-peer platforms, crowdsourcing, and user-generated content. Ultimately, this book outlines a synthesis in this regard for news producers, and it advocates innovation in approach, form, and purpose. This book provides a framework for performing audience-engaged media work of many kinds in our networked, hybrid media environment. The ideas of a "networked press" and "networked news" are still undergoing development, as journalism and journalists' conceptions of their roles have changed.[3] I believe more needs to be said from the journalism side about what an affirmative vision looks like for professional news media in a world of participatory journalism and distributed publishing power.

Purpose and Scope

This book sets out to do two things. It describes changing dynamics relating to news and public knowledge and, correspondingly, it explores a new approach to media work.

In the early Internet days, there was a sense that society might be better off without a bunch of information gatekeepers. In the cartoon version of the mainstream media (MSM) critique, a corporatized media was responsible for most of society's ills. No reasonable person believed this entirely, but there was a kernel of truth to the devout wish to be rid of such compromised authority. News media have been deeply imperfect in numerous ways. Yet the obvious alternative has not solved everything either. Evan Williams, a cofounder of Twitter, told the *New York Times*: "I thought once everybody could speak freely and exchange information and ideas, the world is automatically going to be a better place. I was wrong about that."[4]

There is a dawning realization that we need quality news media institutions now more than ever. This is the resurgence of an old idea, in fact, one that resurfaces whenever liberties are under threat. James Madison once noted that an unfettered press, in allowing the public to freely examine "public characters and measures" and therefore enable "free communication among the people thereon" has a special role in safeguarding liberty. Such free communication, Madison wrote, has "ever been justly deemed the only effectual guardian of every other right."[5]

A premise of this book—one that is not universally accepted—is that professional news media still do matter and will potentially matter even more in the future. By orienting their practice in a different way, journalists may even make a difference concerning issues of societal polarization and fragmentation. "Good journalism for a long time has been successful at solving these problems," notes Matthew Gentzkow, whose path-breaking research on media and partisanship has continued to advance our understanding of news. "Insights that can make journalism better can be helpful to the broader problem of polarization."[6] There is no instant recipe, however. Changes will take time.

My analysis relates primarily to American media, given that the United States is seeing the greatest test to its system among advanced nations. Nearly all countries are dealing with related changes in technology and information flows, and many have responded in different ways: China, for example, by censorship and control; and Russia by using the new tools and ecosystem to spread propaganda and weaponizing the news system against enemies. Europe continues to take a more protectionist stance toward its incumbent media institutions and a more aggressive stance toward Internet platform companies such as Google and Facebook. My hope is that this US-centric discussion has implications far beyond, as more waves of technology and media originating in America spill across the globe—and as knowledge itself becomes increasingly contested across many societies with the rise of digital, decentralized networks.

This book brings together the fields of network science and journalism, perhaps the first time the intersection of the two fields has been considered in an extended treatment. It also draws together research from other adjacent academic disciplines. I aim to bring some of the latest findings from political and computational social science into the conversation as part of the parallel discussion taking place in media circles about the future of news and democracy. Further, I look to make some original empirical contributions based on data I have collected with colleagues. This book's argument is informed by a half dozen surveys I have conducted with various relevant groups—journalists, educators, news publishers and owners, and young news consumers—and informed by analysis of online data. That survey and online data is presented at various points in this book. I also directly interviewed about seventy-five persons involved in journalism, news ownership, network and social science, and the media foundation world, and quotations from some of these interviews are woven into the narrative.

Understanding how people are comprehending the world differently through media forces us, at some level, to discuss matters of what is sometimes called *social epistemology*.[7] This can sound daunting, but I do not refer to the study of the nature of pure knowledge or to a general philosophical theory about true justified belief or human cognition. Rather, following the political philosopher Russell Hardin, I am talking about "ordinary knowledge."[8] This means an account of the "general pattern of individuals' available knowledge" or "pragmatic street-level knowledge" that becomes the basis for beliefs about everything from politics to science.[9] Almost all of our knowledge is socially generated. It does not come through direct experience or experimentation. We get it through testimonies from friends and neighbors. Increasingly, we get it through digital, socially mediated channels and platforms.

Understanding the changing role of news is a vital part of any account of socially generated and subjective knowledge. Such an account must grapple with what I call *social facts*, or media content accompanied and influenced by information indicating social attention or approval. This might be a video with a million views or a comment that has been liked a hundred times. News and information now come accompanied by social cues, which is a kind of fuel for diffusion. Social facts are the visible and invisible body of online public expressions and data traces—the articulation of mass interests and preferences that show up as trending topics, memes, recommended stories, viral videos, and other social media forms. (As a term, "social fact" dates back to the late nineteenth century and the French sociologist Émile Durkheim; my usage here is more specific to digitally and technologically driven phenomena and systems.)[10]

Social facts now suffuse and influence the entire media ecosystem; they are data used to drive what becomes public knowledge. Filtering algorithms and recommendation engines in news aggregators and social media pick up on social facts, steering audiences to content. Social facts in this way are entangled with what people believe to be true and interesting. Even when we get information through traditional broadcast and newspaper media, it has often come to the attention of producers, reporters, and editors through socially mediated channels.

The way that society is changing with respect to public knowledge is frankly quite troubling. These changes help create the context for increasing political polarization. Media fragmentation allows for greater diversity of voices and views, but it diminishes the chances of anchoring the broad public in common, well-justified knowledge. Religious, racial, and geographic identities, as well as issue preferences, have become startlingly and almost mechanically aligned with being a Republican or Democrat, or with left or right ideologies. This is something relatively new in modern society. The political scientist Lilliana Mason has documented this trend toward alignment and concluded that it has deep roots in human psychology.[11] Once group identity is formed, it becomes very difficult to break down.

As we discover more about how much social identity serves to color and constrain epistemology, we should be prompted to consider how journalism might address the underlying wiring of identity—the social networks—rather than just look to increase the sheer volume of factual public information. This is particularly important at this moment, as the process of what Mason calls *social sorting* in the United States has accelerated. Journalists have long talked about the capacity of news to help make democracy work. In a world of expanding social media channels and ICTs, democracy is changing. Journalism must too.

The central argument presented in this book is that news media producers must orient their practice toward fostering what I call *networks of recognition*, dynamic patterns of social connectivity that generate engagement and, ultimately, increase shared public knowledge. Recognition, as I will explain, means something more than mere awareness or attention. It is an idea rooted in a particular conception of democracy and relates to its emerging needs, which are different than those of the past.

Networks of recognition gather their public power by being embedded in the shared concerns of citizens. In the following chapters, I explore how this concept can help organize and reposition news work in the age of networks. This networked concept of news work is the core theoretical perspective presented in this book, a synthesizing framework around which I develop a number of related explorations and insights about media and social networks.

For media practice, the potential for fostering networks of recognition begins in deeper understanding and knowledge on the part of the journalist about citizens' stake in issues. Here I mean knowledge in the more traditional sense of systematic information.[12] Journalists must engage more with data, social science, and research, all of which are instruments for understanding why issues may matter to citizens. Whether we consider monetary policy, patterns of crime, environmental degradation, or teacher quality issues, journalists need knowledge to be able to grasp why issues are relevant to people.

What I am arguing in this respect may seem counterintuitive at first blush. In the face of a fragmenting world, with knowledge fracturing and every niche group selecting only confirming information, I contend that news practice needs to be become much deeper. It must adopt an approach that the media scholar and political scientist Thomas E. Patterson has called *knowledge-based journalism*.[13] This is not because I believe that journalists armed with ivory-tower understanding can suddenly convince everyone of "the facts." That would be naïve. Rather, knowledge is necessary in journalism precisely because it is the key to driving social connectivity; it is the key to relevance and to answering "Why does this matter?" Knowledge is also the key to getting away from the superficial story, the horse race story (who's up, who's down), and the personality-driven media food fight—the stories that leave us with nothing substantive to say, except to argue from a place of conflicting values. In a world of information abundance, journalists are faced with difficult decisions about selection as they look out over a sea of online voices and content. The operative challenge now is selecting what to cover and the information and voices with which to contextualize stories. To choose wisely, we must have knowledge.

In a society increasingly defined by decentralized networks, news producers must understand how to engage networks of citizens in issues. This is achieved by knowing how citizens are mutually connected and allowing them to reflect on those connections. By apprehending key stakes, news producers create the possibility of real public interest, which is its own incentive to learn. News practice becomes about being *generative*, not just informative.[14] This virtuous cycle fuels the co-creation of public knowledge through social media.

The discussion here is relevant to a variety of kinds of media work, although reportorial journalism—with its aspiration toward fairness and dedication to verification—is the chief domain considered. Throughout, I use the terms *journalism* and *journalist* to signify a fairly broad spectrum of news media producers that have a public service or civic dimension. This may include opinion- or even some advocacy-oriented work, as well as traditional, "objective"-oriented work. Preserving the traditional core of journalism is vitally important, but newer forms of journalism that emphasize social responsibility, such as solutions-focused journalism, are adding significant diversity to the media menu. This book speaks to these many diverse media forms.

There is a caveat that I should state up front: the ideas I am advancing may not be for everyone. Many legacy media outlets will undoubtedly stick with old formulas and formats for years to come—and to some extent with good reason. The old mass media strategy has its place. Producing volumes of daily, general-interest news items—a kind of industrial-era, commodity product—still pays the bills for some, although that proposition is increasingly on shaky financial ground. A different media strategy, one that focuses on networks of recognition and knowledge-based, generative approaches, may not be feasible given finite resources. Yet my analysis and argument aim to look over the horizon to the coming decades, when much of the twentieth-century mass broadcast/media industry has faded away or transitioned more fully to digital. The emboldened generation of news nonprofits—ProPublica, Texas Tribune, and the Marshall Project, to name a few—are in many ways better positioned to perform this new kind of networked (and deeper) journalistic work, as are public media and well-endowed national brands such as the *Washington Post* and the *New York Times*.

Although this book does not address the business model problems of media to any great extent, I do believe there is an economic dimension to my argument. Journalism will be valued in the marketplace increasingly to the extent that it is a more knowledge-based, value-added product— something distinct from what Internet companies can serve up programmatically. "On the economic side, journalism has to move from being a commodity profession to a value-added profession," Nicholas Lemann, the former dean of Columbia Journalism School, has remarked. "I sometimes say, half-kidding, that we've operated traditionally on the hunter-gatherer model of journalism. And if we are to have a future as a paid profession, we really have to prove in the age of the Internet that an actual paid reporter or editor does something beyond what somebody just writing comments from their house could do."[15] The economic and social necessity of news moving toward knowledge has enormous implications for media practice and news organizations in the future. The Internet has increasingly been "deskilling" journalists—as the media economist Robert G. Picard has put it—by taking away their comparative advantage in terms of accessing events, persons, and information remote from average citizens.[16]

In any case, networks, knowledge, and the future of news are this book's focus. Its concern is the interplay among ICTs and the evolving discipline of engaging the public on civic issues. Both understanding and harnessing new dynamics and energies are vital for journalism and democracy. These are not short-term dynamics, and they will play out over decades. What I hope is to identify something enduring. The concept of networks of recognition, I hope, will help clarify the goal—the pattern of success—for news producers in a hybrid era of many overlapping media ecosystems.

Chapter 1 of this book articulates the changing civic needs to which news media must respond and outlines key concepts relating to networks of recognition. Chapter 2 looks at some case studies to illuminate how news and information flow through digital networks and create public knowledge. Chapter 3 explores how the structure of knowledge is changing—and how knowledge is being increasingly contested—in a more networked, socially mediated world. Chapters 4 and 5 explore the emerging science of networks, as well as the types of biases embedded in communications architectures, to help us understand new media ecosystems better. These chapters feature deep-dive narratives that surface hidden histories, from tracing how social network theory made its way into the practices of BuzzFeed and other viral publishers to the under-appreciated, and historically consequential, early technical decisions of Facebook's Mark Zuckerberg and the Google founders.

Chapter 6 pivots toward consideration of society's potential news needs long into the future, examining issues such as how artificial intelligence is likely to impact news media and the role of journalists as society's "data locksmiths." Chapter 7 explores how journalists can better prepare for an increasingly data- and knowledge-driven, and networked, future. I argue that transparency in media work and the capacity to communicate uncertainty are crucial. Chapter 8 outlines some of the limits, internal debates, and problems faced by journalism in terms of deeper audience engagement. In chapter 9, I consolidate some of the key research literature on overarching issues such as public trust in news, media polarization, and the effects of diminished news capacity on democracy. That chapter is meant to serve as the more precise documentation for the broad shifts in news and society that my overall argument assumes in earlier chapters. Finally, the book's conclusion provides historical perspective on our media moment and suggests, from a public policy perspective, some ways forward for supporting networked journalism.

Broadly speaking, chapters 1–5 and the conclusion will be of interest to most general readers, whereas chapters 6–9 are more focused on issues internal to journalism practice.

Paradox of the Knowledge Society

In ordinary times, a journalism studies book might take as its assumed audience a small world of academics and practitioners. Wider communities typically regard journalism much as they might the field of accounting or engineering or farm equipment manufacturing: a function necessary for society to run, but of no real compelling concern. There is a jaundiced old newspaper joke that every story ends up wrapping fish in the market the next day. Despite a tendency toward constant self-analysis in the media industry, and a slightly inflated sense of self-importance ("We are the only business mentioned in the Constitution!" say journalism's champions), almost no one else regularly thinks about journalism as a "thing" per se—at least under regular circumstances. Hollywood takes a passing interest every few decades (*All the President's Men* [1976]; *Spotlight* [2015]), heightening attention however fleetingly.[17]

Now questions of journalism and democracy have come to the fore in a remarkable way. We find ourselves probing fundamental issues: How does a democracy operate without commonly agreed-upon bodies of fact? Who is responsible for provisioning those facts? What if the structure of knowledge is increasingly flat, with multiple pathways to understanding? What role should media institutions play in addressing these issues?

We stand at a cultural inflection point in terms of the visibility of information-mediating institutions. Although awareness of "the media" as a substantial force in society dates back several generations now, the awareness has seldom engendered such general discussion and curiosity. Press criticism, once a narrow subfield for specialists and a kitchen-table pastime for citizens, has become general. Striking public polls have continued to show evaluations of news media at all-time lows. In the United States, we are evidently now a nation of press critics.

Further, the increasingly partisan media we consume has made criticism and mockery of media that cover whatever views the individual in question doesn't hold—of media on the "other side"—a staple of Americans' news diets. We see over the fence, and we don't like it much. There is a broader general awareness of the media landscape than ever before. The *third-person effect*, the tendency to believe that others are more affected by messages than they really are, fuels a basic fear that society is careening over a cliff. This creates a vicious cycle of distrust. We fear the worst about what the media "out there" is doing to the minds of fellow citizens. There is strong empirical evidence that partisans on each side of the political divide are becoming much more hostile to one another and that their perceptions of polarization have become stronger.[18]

Some of these trends had been building for decades. How did large portions of society come to loathe the "mainstream media"? Scholars cite the Nixon White House in particular as setting in motion the forces of critique from the ideological right. These were accelerated in the 1980s by new conservative forces, emboldened by the Reagan presidency, that set out to contest the authority of mainstream journalism. These forces are the "deep roots" of President Donald Trump's attacks on the press.[19] A critique from the left, too—centered around news media's reinforcement of existing power structures and its suppression and distortion of marginalized voices—certainly has played its own role. These critiques have been louder and softer at various moments.[20] Every so often, America has had a moment of convulsion over news media.

In *Liberty and the News*, published a century ago, Walter Lippmann wrote that in "an exact sense the present crisis of Western democracy is a crisis in journalism."[21] During and after World War II, the Hutchins Commission on Freedom of the Press was convened to assess the role of news in a democracy and, after several years of debate, recommended that the press, given its vast new powers, adopt a greater sense of social responsibility.[22] In the wake of civil unrest in America's cities and the turbulence of the 1960s, the Kerner Commission weighed in memorably on the issue of race and media, saying American society is "moving toward two societies, one black, one white—separate and unequal." Journalism had neglected its responsibilities: "By and large, news organizations have failed to communicate to both their black and white audiences a sense of the problems America faces and the sources of potential solutions."[23]

The media crisis narrative extended on. In the late 1990s, the sitting vice president, who had been thinking about the problems of mass media since his undergraduate thesis at Harvard (titled "The Impact of Television on the Conduct of the Presidency, 1947–1969"), convened yet another commission tied to the advent of digital broadcasting.[24] That vice president was a former journalist named Al Gore, who set up an advisory committee to seize new trends in media to try to push for reform. On December 18, 1998, exactly one day before the US House voted to impeach President Clinton as a result of the "Lewinsky scandal"—an event that a majority of the public regarded as an absurd, media-driven circus—the Gore Commission issued its findings: it advocated for, among other things, more robust public interest requirements for broadcasters, who should seek to improve the quality of public discourse.[25]

Like clockwork, a decade later there was yet another commission. In 2008–2009, a bipartisan group was convened as the Knight Commission on the Information Needs of Communities in a Democracy. Its members delivered a nuanced message that focused on unequal access to news and information: "The digital age is creating an information and communications renaissance. But it is not serving all Americans and their local communities equally."[26] In 2011, the Federal Communications Commission (FCC) issued a report along similar lines, noting a growing deficit among many regional and local markets as many news institutions had begun to contract.[27] Those two reports were inflected by economic recession, but their focus was not mistrust, misinformation, or politicization per se.

We now have at hand an altogether novel moment of crisis and convulsion, one that compounds previously noted problems with new challenges brought on by social media and political polarization. In contrast to past moments of crisis, much of society is involved—not just elites in news media, universities, government, and foundations. Suddenly people at many levels of society are reflecting back on information itself. This awakening is, in some sense, just a budding awareness of things that have always been. Society has always had social networks; news—including slanted, biased, and outright fabricated "news"—has long been with us. The writer David Foster Wallace began a now-famous speech with a relevant parable: "There are these two young fish swimming along and they happen to meet an older fish swimming the other way, who nods at them and says, 'Morning, boys. How's the water?' And the two young fish swim on for a bit, and then eventually one of them looks over at the other and goes, 'What the hell is water?'"[28] The way that information is being mediated—the water that has always surrounded us—is now increasingly visible.

That our cultural and media moment has become self-aware and "meta," and that the political has suffused all institutions of knowledge and information, is grounded in reasons that go beyond general polarization and partisanship. Technology is forcing and foregrounding the issue of choice. Nearly a half century ago, in a landmark work titled *The Coming of Post-Industrial Society*, the sociologist Daniel Bell presciently forecasted the emergence of this phenomenon more generally, noting that the "knowledgeable society, the technocratic age" will paradoxically become more driven by, and rooted in, expertise, data, and knowledge, while simultaneously more politically contentious. The post-industrial society will likely "involve more politics than ever before, for the very reason that choice becomes conscious and the decision-centers more visible. ... Inasmuch as knowledge and technology have become the central resource of the society, certain political decisions are inescapable."[29]

This paradox characterizes our age. Although awash in an unprecedented flow of data and systematic knowledge, we suddenly find ourselves simultaneously awash in distrust and confusion. Will we ever return to a moment when knowledge is less politicized and the news less polarized? Perhaps not. But examining the relationship between knowledge and the news from first principles, as this book attempts to do, may lead us to important innovations that can help a noisy, diverse, and complex democracy work better.

1 Digital Networks and Democracy's Needs

As we try to imagine where our world of news and digital, decentralized networks may be going, it is useful to meet the future—late millennials and Generation Z, or *Gen Z*. Born mostly in the mid-1980s and later, these persons do not remember a time before the web and the Internet. Their childhoods and teen years have been filled with digital and social media. These groups, particularly Gen Z, are now larger and more diverse than older generations. Their economic power is rising, and they have distinctive behaviors around, and views about, news.

In 2018, a group of researchers I was involved with through Project Information Literacy began a large-scale analysis of the emerging generation's information-seeking and news consumption habits. We surveyed thousands of college students across the country, evaluating their news-, knowledge-, and social-media-related habits and perspectives. We also asked for their Twitter handles so that we could observe their behavior in the online world.[1] We ended up with more than five thousand respondents from a geographically varied group of colleges and universities, in liberal- and conservative-leaning states alike. They were asked about their encounters with news over the prior week.

And so, in the spirit of meeting the future, I want to introduce you to Janice, Tim, Sandra, and Marcus, whose names I have changed to protect privacy. Their profiles help frame our discussion and exploration in this chapter. All four of these young news consumers use social media to engage with and share news. We will get to the survey's overall findings in a moment, but let's first look at just how complex the lives of these highly networked young news consumers can be.

Janice is in her late teens and at a school in the Midwest. She's interested in science. She tweets several times a week and gets a lot of news every day on Snapchat, but she's not on Facebook. If forced to choose from among a group of big media outlets, Janice prefers Fox News, yet the content that she shares is generally more left-leaning. Interested in matters of race, Janice often retweets the account of AJ+ (@ajplus), which labels itself as "news for the connected generation sharing human struggles and challenging the status quo." The reason she shares news is to try to change the views of her friends and followers. She strongly agrees that news consists of objective reporting of facts, but she does not necessarily trust news from traditional sources produced by professional journalists more than news she finds on social media sites.

Tim, too, is in his late teens and is pursuing a business degree at a large, noncoastal state school. A self-described political moderate, he gets a lot of news from YouTube and tweets several times a week about sports, memes, and pop culture references (many about *The Office*). He retweets news quite often from accounts like that of the Associated Press; he has challenged the *New York Times* with a tweet and retweeted President Trump critically. He does not necessarily trust traditional sources more than social media.

Sandra is a liberal in her early twenties, preparing to go from a midwestern school to a graduate school on one of the coasts. She gets a lot of her news from Twitter and Snapchat. Her posts are sometimes about her personal activity; often they are about pop culture or celebrities and memes, as well as about domestic and international news issues. Her topics range from the Second Amendment and school shootings to Syria and climate change. She has retweeted both President Obama and President Trump (criticizing the latter). Sandra believes sharing news gives her a voice about larger causes in the world. Asked if she trusts news from traditional sources more than social media, she agrees "somewhat."

Marcus is a conservative from a very Republican-leaning state. He sees news about once a day on Facebook. An engineering major, he tweets on a weekly basis, sharing a lot of conservative content opposing gun control, prompted in large part by the controversy over the Parkland, Florida, school shooting in February 2018. He also shares humorous media content. He has approvingly retweeted President Trump regarding the need to show military strength against threats from North Korea, and Marcus has tweeted about bias in the news and about veterans and armed services. He does not necessarily trust traditional news sources more than news he finds on social media, but he is neutral on the question of whether or not journalists deliberately insert their bias into stories. Marcus thinks that journalists make mistakes but generally try to get their news stories right. And he shares news to have a voice and shape the views of his peers.

These four late millennials/Gen Z'ers, taken together, give us an intriguing snapshot of an emerging generation. What do we learn? News and social media are bound up inextricably for them. They may have strong political views, but they are not always predictable. The way that they assign trust and assess credibility does not always correspond with traditional notions of authority. Sharing, commenting on, and engaging with news is part of the social identity of these young people; these practices are vital to their sense of agency and empowerment in shaping the world in which they live. News in some ways is a "platform" on which they are able to operate. It is the generative fuel and the raw material for their participation in the public sphere. Janice, Tim, Sandra, and Marcus use news to build public knowledge within their online communities.

In the overall Project Information Literacy survey, the number of young people who said they got news through social media more than once over the past week was 83 percent, with 72 percent saying they did so at least once a day and more than half saying they saw news on social media several times a day. By contrast, only 5 percent of respondents said they got news at least once a day through print newspapers or magazines and 15 percent did so through television.

In terms of news consumption on social media by specific platform, about 70 percent of respondents said they had seen news through Facebook at least once over the past week; about 56 percent said the same of encountering news through Snapchat. Those same frequencies were 54 percent for YouTube, 52 percent for Instagram, and 42 percent for Twitter. Nearly half of respondents said they got news from peers (either online or face-to-face) at least once a day. More than half said they consumed political news once a day. About a third of the young people surveyed said they had shared political news on social media over the past week, and about a third had shared a "political meme," or a humorous image, video, or text-based message.

Why do they share things online? About half of respondents said that sharing news lets their friends and followers know about something important, and nearly half said sharing news gives them a voice about a larger cause in the world. About a third of young people said they shared news to try to change the views of friends and followers.

We also asked this group of late millennials and Gen Z'ers about defini-tions of news, with various questions that probed different issues. Some 83 percent agreed that news is information that is useful to their life. About a third of respondents did not agree that news consists of objective reporting and facts. More than two-thirds said that the sheer amount of news was overwhelming on any given day, and 45 percent agreed, to some extent, with the proposition that it was difficult to tell real news from fake news. Finally, about half of respondents either "strongly" or "somewhat" agreed that journalists deliberately insert their opinion into stories, while only 42 percent said they strongly agreed that they trust news more from tradi-tional sources produced by professional sources than news found on "social media sites where anyone can post news."

Ultimately, the survey data we gathered reveals something about what it is like to have lived in a nearly always-on, Internet-connected society, in which news is not about traditional, appointment-driven habits (the morn-ing paper, the evening news). It's a life in which news often finds you, not the other way around. For the purposes of this book's discussion, these por-traits raise the important question of what news will be when these young people come fully of age, start families, buy houses, and settle into the heart of the workforce. It is an urgent question, as this wave is coming soon. Surely, some of their more fluid information-access practices will harden into habits; certain news brands may become more constant in their lives. But the one thing we can likely be sure of is that a world of networked, socially mediated news and information will create the context for what-ever habits they may settle into. And that alone will represent a sea change.

We can take our speculation to an even higher level. The philosopher of information Luciano Floridi has asserted that the rise of billions of ICTs, and their integration with our lives, will eventually transform reality into what he calls the "infosphere." As he puts it, "the suggestion is that is that what is real is informational and what is informational is real."[2] Enormous portions of our waking lives will be spent in digital space (indeed, they already are), with everything from money and transportation to job-finding and dating all mediated by information platforms. Identity is formed and becomes substantially bound up with one's digital presences. Notions of news, its role, and its function, will change. "We are witnessing an epochal, unprecedented migration of humanity from its Newtonian, physical space to the infosphere itself as its new environment, not least because the latter is absorbing the former," Floridi has written. "As digital immigrants, like Generation X and Generation Y, are replaced by digital natives, like Generation Z, the latter will come to recognize no fundamental difference between the infosphere and the physical world, only a change in perspective."[3]

Evolution of News

It's helpful to start from first principles as we attempt to project forward into the future of news and its role in society. There is, as mentioned, increasing confusion over the very definition of what we're talking about: What, after all, is news? This is no easy question for the late millennials and Gen Z'ers we met—nor is it for anyone these days—and it may not get easier in the future.

News is not information; it is a "form of culture."[4] It is a mode of human communication extending out from our most basic tools, speaking and listening, to kindred forms of mass communication such as publishing, broadcasting, tweeting, posting, and digitally sharing. Its meaning and form are embedded in a particular time and place. News is in some sense a form of life, a way of constructing a "symbolic world that has a kind of priority, a certificate of legitimate importance," as the media sociologist Michael Schudson has noted.[5] Its logic, its intelligibility, is therefore contingent, bounded by and explained in history. In this regard, Schudson points to the anthropologist Clifford Geertz and his idea that "culture is not a power, something to which social events, behaviors, institutions, or processes can be causally attributed; it is a context, something within which they can be intelligibly—that is, thickly—described."[6] To describe what news is is to map a whole matrix of variables, from politics and economics to education and the needs of everyday life.

Culturally specific context, then, defines news and generates its meaning. In each era, news has changed in form and content—and purpose. The shipping news and massive tables of printed stock prices are no longer with us. Likewise, news of civil rights violations or human rights abuses or environmental harms, as such, are nowhere to be found in nineteenth-century newspapers. No such cultural concepts yet existed to define news categories or provide a symbolic context for discussion.

Cultural space, in combination with technology, also defines the way news is accessed, delivered, and understood to be relevant to everyday experience. American culture once had newsreels before feature films in movie houses, until the television separated these media and entertainment channels. We now have the shareable GIF and the SMS text alert, allowing instant bits of parody and microscopic nuggets of breaking news to reach individuals. But perhaps these are with us for a fleeting moment. The email newsletter and the podcast were once declared dead artifacts of the early web, but the newsletter has come roaring back as information abundance has become overwhelming, as has the podcast in a technology-driven world that requires ever more multitasking. The journalism historian Christopher B. Daly has articulated a quasi-biological notion of journalism's evolution over time. Journalism institutions see stresses and pressures from the environment, and they live and die as a function of business-related, cultural, technological, and policy-related forces.[7]

For a particular form of information to be grasped as "news," human communities must validate it as such. In this way, news is always a changing and unstable cultural genre, constantly susceptible to cultural evolution. Religious discourse, forms of advertising, conventions of exchange between lovers—all of these modes of communication, these forms of culture, differ radically over decades and centuries. So too with news.

From the European coffee houses of the Enlightenment to the public houses and printing press shops of the American Revolution, the idea of news arose as a way of learning more about the world than an individual could ascertain from his or her immediate network of family and friends. Travelers sometimes wrote down events on public house and tavern ledgers to convey what had taken place outside the community. Distance and geography played a strong role in shaping what was then considered news. The limitations of place created the opportunities for its development and eventual institutionalization.

Newspapers at the time of the American Revolutionary War were partisan in nature, an accepted pattern of structural "bias" that persisted well into the nineteenth century. Researchers have estimated that the percentage of political newspapers that claimed to be independent rose from 11 percent in 1870 to 62 percent in 1920.[8] In fact, the idea of bias might have sounded very odd to citizens of that earlier time, for bias implies an alternative form of journalism that had yet to be invented. Citizens wanted information from the political parties, and printers provided it. In the eighteenth and nineteenth centuries, a great deal of "news" items printed on broadsheet paper were simply reprinted speeches by public persons. There was no other way to know what remote persons of influence were saying or thinking. For decades, early American newspapers also reprinted a massive amount of relatively trivial European news; the cultural ties were still strong across the Atlantic. Useful facts and trends relating to local commerce were also fitted into the inches of news columns. News as a cultural form fit the expectations and needs of those mostly rural citizens—farmers, craftsmen and women, and small-time tradespeople—of the time.

The interview, now a staple of nearly all forms of media, was not invented as a news-gathering method until later in the nineteenth century. Many scholars date the establishment of the technique to Horace Greeley's interview of Brigham Young in 1859.[9] It may come as a surprise to many people who now lament the partisan bile of cable news shows and Internet sites, but the idea of a disinterested, professional reportorial press, untethered from political patronage, is in some ways a historical anomaly; it is a relatively recent invention that accelerated in the post-World War II era.

Scholars have many theories for why, culturally speaking, there arose a need and acceptance for this type of "objective" news. One powerful explanation is economics. The rise of advertising in the late nineteenth century allowed newspapers to begin to move away from the political parties as their chief source of income. Further, as the telegraph began to change the way news circulated, the Associated Press in particular needed to create stories and information that could be run by newspapers of all partisan stripes. Mass broadcast technologies influenced the character of news in similar ways.

More nominally neutral publications and broadcasts allowed for advertising of all kinds to reach the broadest possible audiences. The rise of radio and television, and their potential to reach markets, was the driving engine of this new culture of news. As the journalist and historian David Halberstam noted in his classic exploration of mid-twentieth-century news culture, *The Powers that Be*, broadcasting is the "most powerful instrument in the world for merchandising soap, and it is potentially the most powerful instrument in the world for public service, and it has always been caught between the duality of its roles."[10] Indeed, the neutral public service approach and the broad public audience it gathered helped sell soap to millions.

Policy and law also shape what a society considers to be legitimate news. In the 1500s, English authorities would proclaim any criticism of magnates of the realm as "false news," or *scandalum magnatum*, punishable as treason and heresy.[11] Even if true, such information was not validated as legitimate public communication, or news. In 1690, the inaugural issue of the first American newspaper, *Publick Occurrences*, appeared. The publication was subsequently suppressed by Colonial authorities, however, for treading in dangerous territory. This was not so for the reporting centuries later of the likes of Bob Woodward and Carl Bernstein, the *Washington Post* reporters who helped expose the Watergate scandal that took down President Nixon in 1974, or for the myriad reporters breaking news of scandal and palace intrigue about the administration of President Donald J. Trump. Parameters of law and policy shape what can be said, what will be received, and what will be rewarded. Even now, in the twenty-first century, substantial libel laws more severely limit what can be published as news in liberal countries such as the United Kingdom, in comparison to the broad (but not infinite) protections in the United States for free speech and the press.

In an older era, as now, the report of a coming storm or the swoon of market prices would be highly newsworthy; yet in the past only news media outlets could make the public aware of such events or trends. Now, with Internet-enabled communications technologies, such information is blending into the architecture of digital applications, social platforms, and automated programming interfaces or data pipes. Things that were once "news" proper are becoming just information—a commodity that has less value in the marketplace, as it is no longer in any sense exclusive. The boundaries of news may be expanded or contracted because of technology, as well as economics and law. Technology creates and serves as a platform for information and news.

This leads to a final point that will be discussed in more detail in chapter 6, which explores the role of artificial intelligence in media. No matter how much technology manages to supplant various aspects of the news industry's current functions and job tasks, humans will continue to have a demand for new such layers of some new cultural form called "news." Why? The answer is that humans are meaning-making creatures. "We generate and use meaning a bit like the larvae of the mulberry silkworm produce and use silk," Floridi notes.[12] As a form of culture—and not just "information"—news is fundamentally about meaning-making, a kind of biological function, in the lives of people. We can see this, for example, in the practices and worldviews of Janice, Tim, Sandra, and Marcus, the

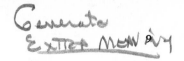

prototypical late millennials/Gen Z'ers we met earlier. Whatever form of news they encounter, more meaning will be created on top of it. They will always create "silk." It is part of who they are.

This reality of human nature does not dictate that a large profession will always be dedicated to the making of news (that depends on business conditions and marketplace creativity), but it does suggest that the demand from the public will be there persistently, climbing up the value chain for ever more and different forms of meaning, even as machines and computation continue to supplant lower information-production functions. "Culture is the dark matter of the social universe, invisible but exercising extraordinary power," social theorist Jeffrey C. Alexander observes. "The meanings of journalism are fervently formed and fiercely delineated, and the cultural power of the profession resists technological and economic determinism."[13]

From Authority to Authenticity

So news is a form of culture. This fact should necessarily prompt us to think about how news can align better with new cultural expectations and evolving needs. I want to first consider these cultural shifts in a bit more detail before then exploring and defining what the specific responses and innovations of media producers might be.

Media of all forms provide pictures of society, of its disruptions, wrinkles, novelties, and concerns. All media are samples of society, in some sense, with varying scope, subjectivity, and layers of mediation, both individual and institutional. An eyewitness photo attempts to be a direct sample of reality. It is "mediated" only by an individual and his or her point of view. Professional news stories are frequently less direct samples, from larger slices of reality. They are carefully packaged and institutionally mediated, insofar as they are embedded in professional norms. Fictional accounts and false news are barely samples of any reality at all and of course are highly mediated samples at that. Media thus exist on this continuum of narrow and wide sampling and types of mediation. Although individually mediated pictures may have just as much (if not more) bias embedded in them, they almost always appear to be a more "direct" sample of reality.

The amount of individual sampling of reality available for consumption has grown exponentially with the rise of Internet platforms that empower citizens to produce media, as well as vast automated systems that collect data, images, and media of many kinds. Each day, an estimated two billion digital images are posted on social media, and more than one billion hours of video are watched globally on YouTube alone.[14] YouTube itself is, arguably, the largest repository of human culture—at least in its common or vernacular form—ever produced. With a billion users, YouTube is a "fairly representative portrait of human behavior," argues Kevin Allocca, the platform's head of culture and trends.[15]

The ratio of samples of individually mediated reality (user-generated, by the world's citizens) compared to institutionally mediated samples (professional news and products of other culture industries) has changed in breathtaking ways over about two decades. These titanic shifts are fueling a hunger for individually mediated realities and for content purporting to be direct and "undermediated." There is what we might describe as a *hunger for authenticity* and a move away from the veneer of authority. Media issuing from individual experience and expression are gaining traction. The ubiquity of such content changes habits and consciousness. It reshapes how people recognize what is interesting and what they take to be true. The very notion of institutional mediation—of powerful groups conferring the authority of the real, sanctioning representations as true—is under strain.

News media, as much as any other mass cultural form, are feeling these changes. Our knowledge of the world around us was once dependent on news institutions, which constituted hierarchical broadcast networks. This is no longer true. News organizations once furnished the proverbial "first draft of history." Now, through more horizontal, distributed networks, we have other forms of primary evidence of what just happened, allowing for new narratives and alternative ways of knowing. The trend of expanding content from non-institutionally mediated sources will only accelerate. It is a major factor in the structural crisis of news.

The public's sense that something is wrong with news media is also a result of overpromising and underdelivering. The era of President Donald Trump has underscored this point, although its origins long predate his election. News media are expected to provide the primary accountability or watchdog function for the government and other centers of power. Yet the issue of collective legitimacy is always tenuous and unresolved when it is left to private media institutions to perform public watchdog functions. It is almost always better for larger, more accountable public institutions (oversight committees, attorneys general offices, regulatory agencies, etc.) to perform such critical functions.

Journalism, now much poorer and smaller relative to its more robust industry past, can do no better than catch occasional malfeasance. Truth be told, it has never had systematic power. Its power is in providing high-impact examples of wrongdoing. It cannot police democracy in any general way, and the expectation it will do so leads to disappointment. Investigative journalism remains important and vital, but it cannot pass laws or elect virtuous people after scandal has been exposed. It will not necessarily make things "better"; society must do that. We expect too much of journalism in this regard.

Further, it is clear that we are asking news media to play far too big a role in elections. Because US party primaries are now essentially a "media primary"—US primary elections once relied on convention delegates but now rely on open state voting—news media are put in the tenuous position of picking winners and losers for the voting public by focusing attention on some candidates while excluding others. They do this with increasing negativity, trapped in a pressure cooker of ratings wars and online metrics.[16] Journalists are neither prepared nor qualified to play this role of deciding elections. Again, the situation only fuels public disappointment and cynicism.

Any honest conversation about the news future should begin with a discussion of limits. There are certain problems that must be solved by institutions beyond journalism—by a renewal of the public sector, of the electoral system, of political ethics. Journalists cannot reprogram society. "If ... modern societies are undergoing silent value changes that lead to a retreat from the public sphere then all efforts to redefine the journalistic role are of no purpose," media scholar Wolfgang Donsbach observes. "The solutions to this are far beyond the control of professional communicators and their educators and need to be addressed by general education."[17]

The public's encounters with news are increasingly taking place through digital social networks. Of course, social networks have always existed in human societies. The difference now is the speed and scale of networked communication, as well as the ability to maintain loose ties with many, many persons across space and time. Our cultural moment is also marked by the phenomenon of news being able to find individuals in a very targeted fashion, not just vice versa. Algorithms, web applications, and automated systems fuel and shape networked communication in novel ways. The interdependent nature of the world is becoming evident through these digital networks, and this interdependence is being accelerated by the associated digital wiring and rewiring. Social networks are becoming inseparable from news: its creation, spread, meaning, and credibility.

Philosophers and social scientists have long noted the socially constructed, socially bounded nature of knowledge.[18] It is now becoming clear just how social all knowledge, even scientific knowledge, may be. Earlier I mentioned the concept of social epistemology, which tries to come to some account of subjective individual knowledge—your knowledge, my knowledge, his or her knowledge.[19] But more generally, we might also consider the socially constructed pathways that lead us to this knowledge. In this regard, science and technology scholar Sheila Jasanoff articulates the concept of *civic epistemologies*, the unique pathways to understanding and shared knowledge that nations and communities create.[20] Cultures develop "shared approaches to sense-making" that dictate what is accepted as public knowledge and what is excluded or doubted.[21] The web, with all of its contending groups, makes possible abundant pathways and approaches. Civic epistemologies may proliferate in any pluralistic society. A society fueled by digital networks accelerates the proliferation.

Recognition and Networks

In the age of digital networks, the unexpected seems an everyday occurrence. From the welter of digital noise arises a biting meme, a viral mobile video, a grassroots network, or a whiplash-inducing hashtag. Bubbling up from a near-infinite pool of ones and zeros emerges a burst-like pattern.[22] It washes over the moment, shifting the terrain of knowledge and modifying the geometry of news. A surprised public frantically responds, but not without a quizzical backward glance. *How in the world did that just happen?*

The rending of the epistemological order as we have known it is almost palpable, registering sometimes as a kind of intellectual dizziness for the public. The cultural changes appear to reverberate down to some elemental substructure of reality, to the quivering atom and the transmitted electron. The combination of changes in technology and society produce something fundamentally new and emergent.

As our review of the lives of young news consumers suggests, news will increasingly derive its power and energy from social context, engagement, and public participation. Digital media scholar Adrienne Russell notes that news has seen a "shift from personalization to socialization" as the web has become more about peer-to-peer platforms. The participation of engaged "networked publics" offers "more plurality in the news landscape."[23] The facts in news stories are socially filtered and mediated, remixed, shared, and commented upon, with citizens adding new knowledge to that generated by journalists. Views and perspectives expressed online, and the news people share, are themselves facts in the world. They are social facts. An online public opinion poll is, in effect, ongoing on every conceivable subject, story, and narrative line. Likes and retweets increase; topics trend and gather comments and shares. Attention itself is a form of validation, shaping our sense of reality. Social facts are part of the cultural miasma through which public knowledge generated by journalism and other media industries will swim, grow, and evolve.

To the extent that news media will be powerful in the future, it will be because journalism centers its energies around the domain of social connection. Journalism must pivot away from a hierarchical model of trying to "inform" individual citizens—a model that comes from a particular conception developed in the mass-media and broadcast era of the twentieth century—to one of trying creatively and knowledgably to connect people and groups. As mentioned, it means being generative, rather than merely informative. I do not mean "to connect" in the vague sense used by the techno-optimists of Silicon Valley. I mean it in the more strenuous sense of to foster recognition. The term *recognition* has several meanings in this context: grasping or apprehending a reality or state of affairs; accepting the validity of others' experience; and acknowledging the mutuality implicit in citizenship within a shared democratic society.

Recognition as a political concept, as philosopher Charles Taylor writes, is a "vital human need."[24] If there has been a central political value guiding the postwar, postcolonial era, it is the principle of equality of rights, the idea that all persons are entitled to recognition and due the same opportunities and protections. Yet I do not mean recognition only in this sense. Although political recognition is a noble aspiration for journalism to continue to help facilitate (review the Pulitzer Prize winners every year to see how much this is regularly done, giving "voice to the voiceless"), what I mean by recognition here points to something broader and more elemental in democratic culture.[25]

Writing about the continuing struggle to resolve questions of race in America, the political philosopher Danielle Allen has noted that we live, whether we recognize it or not, in "networks of gain and loss" as fellow citizens in a democracy. Collective decisions will always advantage some but not others. It is our obligation, she says, to acknowledge the "interrelatedness of citizens" and engender "networks of trust," even in the face of inevitable loss and disagreement.[26] The tangle of ties that constitute society must be acknowledged, renewed, and made visible. As a culture, we are losing this language of trust and the ability to deal with loss. This loss is manifest in the extreme partisanship seen across society. In part this is happening because mutual ties and connections are not recognized and reflected upon.

On many issues, our culture no longer lacks sufficient information. Instead, we frequently lack the associative ties to facilitate the knowledge necessary for recognition. A central doctrine—indeed, mythology—of the press has been that it *informs* citizens who, implicitly, have some deficit in information. This model needs to be revised. In fact, scholars have been calling for a revision of journalism's informal "theory of democracy" for some time now. The media sociologist Herbert Gans has asserted that a journalist's self-appointed role of informing citizens suffers because journalists lack sufficient knowledge to make judgments about which information should be selected in the first place; they also lack awareness of how their news agenda may serve the economic interests of media, not citizens.[27]

News media will best serve twenty-first-century society and democracy not by seeking to generate ever-more information but by working to foster what we might call *networks of recognition*, allowing disparate citizens to become aware of and reflect on their common ties and concerns—and thereby to sustain collaborative, interpretive activity. This aspect of reflection on connections is vitally important. The goal of journalistic practice therefore is not mere information about the world but knowledge of how citizens are connected to the world.

Journalists increasingly cannot impose agendas. Instead, networks of recognition form and can be fueled when conditions of ambient readiness for dialogue already exist among citizens. The "informing theory" of journalism thus gives way to more of a generative model focused on social connection. In the language of network science, the journalist is an *edge* (links or ties) generator, producing connections to *nodes* (persons or entities), enriching and highlighting the often-invisible edges in society to produce networks of recognition. For journalism practice, this sometimes may mean engagement with third-party communications networks such as Twitter and Facebook, but the idea runs deeper than just conducting any particular digitally networked activity. Investigations and other in-depth forms of journalistic activity also may help foster networks of recognition.

To foster true recognition, journalists must have knowledge deep enough to know citizens' shared stake in issues. By articulating these shared stakes and fostering recognition, journalists ensure public knowledge formation will take place and networks will form. Citizens thereby recognize themselves as community members through several mechanisms. Previously latent convictions and unarticulated beliefs, perhaps already widely held, may be revealed collectively; in addition, novel facts and narratives may help new collective opinion formation.

There is a strong connection, often unremarked upon, between knowledge and networks in the context of journalism. They can feed into a kind of virtuous circle. Journalists must be able to grasp and articulate the stakes for citizens in issues. Increasingly, for citizens to engage in issues that have complexity, they need a basic level of understanding to organize and coordinate collective action. Often this means an awareness of common problems. The more engaged citizens are, the more they will be informed, as interest in issues is an incentive to learn. By knowing more, journalists are therefore more likely to engage the citizen. They are more likely to create networks of recognition.

Consider the context of megastories from recent years about police use of force in Ferguson, Missouri, and the water quality crisis in Flint, Michigan. Police abuse was rampant in the former case, lead poisoning in the latter. Signs of these communities crying out for help had long been available. Indeed, many other communities had similar problems. Citizens had complained, but few had listened, including news media institutions outside of local areas.

Consider, too, the ongoing economic and social troubles of so many noncoastal towns in the United States: the scourge of opioid addiction and closed mainstreet businesses, a sense of being left behind. It was only after the 2016 election that news media realized a major story they had missed almost entirely. And then in late 2017 there came the astonishing revelations of sexual harassment and abuse, galvanized by the #MeToo campaign, from a vast network of women who had long stayed isolated. Again, the stories were right there in front of us. The same was true with the pervasive pattern and growing trend of housing evictions across the United States, which drives many struggling families into further poverty traps. Only after sociologist Matthew Desmond connected the stories of evicted families and began mining court records did the country begin to awaken to this profound problem of housing insecurity, driven in part by inadequate public policy.[28]

When networks of recognition formed around all these issues—often a dynamic product of research, news media reporting, and social media activism—it all became appallingly obvious: there had been a failure to recognize structural problems in society.

The operative challenge for journalism is becoming clear: to help organize attention in a knowledge-based fashion. When networks of recognition snap into place, the world is better illuminated. The social capital that democracy needs may be generated. Networks of recognition are patterns of social connectivity among disparate persons and groups that create "bursts" of public knowledge. Recognition leads to revelation. New revelations generate more recognition and awakening, increasing the size of the network. Even in the absence of complete factual agreement, networks of recognition bring together persons with common concerns, involving them in shared deliberation. Journalists can help sustain these networks and this deliberation by continuously guiding attention. But first they must work to detect problems of systematic societal failure, such as the police-involved shootings or sexual harassment examples, more efficiently.

How can the next Fergusons and Flints, the quiet social scourges and the buried secrets, be more quickly spotted, analyzed, and elevated to the mass attention of policymakers and a motivated public? How can community voices that were long ignored in each case be knitted together in an online public more quickly? The more journalism seeks to answer these questions, the more it will create tremendous value, ensuring its place in a changing society with changing news needs. This must become one of journalism's core functions if it is to thrive in the future. This is not to recommend a journalism of advocacy but rather a journalism of connection.

As communication theorist James W. Carey suggests, "About democratic institutions, about the way of life of democracy, journalists are not permitted to be indifferent, nonpartisan, or objective. It is their one compulsory passion, for it forms the ground condition of their practice."[29]

Social Realities of News

To understand how the concept of networks of recognition fits in with this structural crisis in news, we might entertain a thought experiment: Imagine that all professional news institutions disappeared tomorrow. We would be prompted to ask two questions under such a counterfactual. What would evolve? And what *should* evolve? Old-style media institutions might initially grow back, but that would be only a temporary phenomenon. The old order would grow back, if at all, against tremendous friction and challenge. We can see that with the rising generation of news consumers—with our late millennial/Gen Z'ers Janice, Tim, Sandra, and Marcus. The continuing existence of old methods and approaches to creating and delivering news hinges so much on slowly dying news audience habits ingrained over past generations. This is sometimes called *appointment news*—the morning paper, the evening news, the drivetime news show. We are moving from a news world defined by appointment and habit to one defined by context, with stories reaching us around the clock as a function of their relevance within our digital social networks. It is a world of user-generated context, not big-media-driven appointment.

Given this shift, we might ask: What would be ideal? What *should* evolve? One response might be to let the democratized, citizen-centric model take hold. Get rid of "the media." Just let the gatekeepers die—and good riddance. This view, or a version of it, was a fashionable response among intellectuals as the social web and participatory media began to take off in the early 2000s. Yet all at once there is an awareness that news institutions may be needed more than ever. As the philosopher John Dewey noted nearly a century ago now, communications technologies are insufficient to create a public, which will remain "shadowy and formless" without what he called "social inquiry"—data-gathering-oriented activity that is the basis for new language and symbols, which can generate common meaning. Journalism can play exactly this role.[30] Publics, in Dewey's view, are in large part about knowledge formation.[31] Growing cycles of public knowledge creation around issues create publics, or networks in which members recognize their shared stake.

In the still-emerging world of the social web, a glimmer of hope arises for the enduring role of professional news media. The fundamental laws of networks actually favor journalism in important ways. Networks typically look hub-like, with certain nodes central to guiding attention. Network science assigns a "degree" score to nodes in networks, calculated by the number of links attached to a given node. Journalists and their institutions are "high-degree" nodes. As this book will explore in detail, the science of networks clarifies the ongoing need for high-quality journalism. Although news media will not regain the social centrality they had in the broadcast era, in a networked era they play an outsized role in directing attention, which is the scarcest and most precious commodity in the digital world. Rather than "gatekeeping" as media did in the broadcast era of the twentieth century, news media in the future have the capacity to help networks form. This decisive role in shaping attention has to do with how high-degree nodes influence the operation of networks, and certain mathematical and sociological rules come into play.

One of the largest studies of social media ever done—more than one billion links on Twitter—suggests that attention online is almost always the result of news media paying attention to a given topic.[32] Likewise, a monumental study published in 2017 in the journal *Science*—in which researchers partnered with dozens of news organizations across the country to conduct an extraordinary randomized control trial—found that publication of news items on many topics led to a 63 percent increase in discussion of the topic online and on social media. "Given the tremendous power of media outlets to set the agenda for public discussion, the ideological and policy perspectives of those who own media outlets have considerable importance for the nature of American democracy and public policy," note the study's authors, Gary King, Benjamin Schneer, and Ariel White.[33]

What should astonish almost all of the Internet's early theorists is the central role that news media continue to play within decentralized social platforms and networks. Research continues to show that news media are responsible for making visible important information, even within social platforms. "We are not seeing the end of hierarchy," Duncan Watts, a computational social scientist and principal researcher at Microsoft Research, has noted. "We may be seeing the replacement of one hierarchy with another hierarchy. We may be seeing the replacement of one set of gatekeepers with another set of gatekeepers. … But we're certainly not seeing an egalitarian world where everything has the same chance to become known or accessible."[34]

As the communication and media scholar James Webster suggests, early expectations of a world of participatory media that could replace hierarchical mainstream media has not been borne out by the data; the "evidence suggests the emergence of what is more aptly labeled a massively overlapping culture."[35] Instead of a media world that looks flat and fully democratized, we might expect a messy "commonality to cultural consumption"—a concentration of public attention around certain topics and forms of discourse, even if people encounter them through myriad, fragmented channels and outlets. "I think the cultural ballast provided by old media will remain with us," Webster notes.[36]

News media, as high-degree nodes, will have an important place even in a world of near-infinite user-generated content and vast distributed networks. Part of their future role will relate to creating valuable civic space itself, requiring news practitioners to create new forms of autonomy in technical systems that do not always share their public interest values. The journalism and communication scholar Mike Ananny has articulated a new concept of press freedom in the networked ecosystem, which involves ensuring the right of the public to learn about and express views on diverse issues and concerns. "Instead of seeing their profession as the production of information—faster, more immersive, more shocking—journalists might see their primary responsibility as the creation of environments for listening," Annany asserts.[37]

The pressing question for news practice then becomes whether or not journalists and their institutions are adequate to the task of knowledgeably and usefully creating such conversational environments—and thereby succeeding in this new world. Moreover, the reasoning and deliberative process that unfolds online is often suboptimal, and journalism must bring knowledge to improve public deliberation. Communication scholar James S. Ettema notes that "journalism cannot be content to passively transcribe that reasoning or to uncritically preside over a forum for its presentation." Rather, journalists must be "reasoning participants."[38] For practitioners, deeper knowledge and a grasp of networks and their shaping architectures are the keys to fulfilling an obligation to increase the quality of delibera-tion. News media must navigate the "post-truth" age by building deeper subject-specific knowledge, as well as greater understanding of the con-structed nature and biases of networked communications. If this can be done, journalism can meet the challenges of the structural crisis at hand.

To foster valuable forms of recognition, journalists will require much greater knowledge about policy and issues, social science and data. It is apt that the roots of the word itself, *re-cognition*, have embedded in them the idea of deeper thought, of reexamination through the application of the intellect. A networked media world requires that journalists become steeped in the knowledge necessary for them to select what's important, to guide the attention of networks in constructive ways that produce value for citizens. Selection becomes all-important—and selection without knowl-edge is just haphazardly throwing darts, hitting bogus trends and story-lines that mislead and misinform. Above all, journalists must understand networks themselves, the new ecosystems and structures in which news practitioners will operate and listen, engage and connect.

News practice increasingly will be conducted in constantly iterating and unfolding social processes in fast-moving, participatory networks. Journal-ists will need to grapple with online communities that produce their own versions of truth, generated and influenced by social facts; news itself is becoming increasingly socially mediated. This social reality of journalism must be accounted for, reorienting the mission and goals of the practice.

Democracy's Network Deficit

How can media practitioners, employing a new approach, help society grap-ple with the underlying problems of American democracy? The troubles are many. Electoral maps show ever-sharper geographic divisions among parti-sans, fueled by what political scientists Steven Levitsky and Daniel Ziblatt call an "existential conflict over race and culture."[39] Geographic sorting by cultural affinity has continued, leaving large gaps between the values and policy preferences of rural and urban citizens. "If one thing is clear from studying breakdowns throughout history, it's that extreme polarization can kill democracies," Levitsky and Ziblatt note in their ominously titled book *How Democracies Die.*[40]

Meanwhile, the fraying of the local civic fabric in the United States has been traced to a contraction in civic association membership, across a wide variety of organizations in different domains, and to even less meaningful contact with neighbors.[41] The strong local grassroots organizations that once anchored movements advocating for political change have frequently been superseded by national lobbying organizations. The collective efficacy of local groups has diminished across the nation. "Yawning gaps have opened between local voluntary efforts and the professional advocates and grant makers who seek national influence," political scientist Theda Skocpol notes. "As parallel changes of this sort have unfolded in both electoral and associational life, American public discussions have become polarized in superficial rather than consequential ways. And public policymaking has tilted upward, even in an era of growing social inequality."[42] Sociologists have cataloged a broad trend away from civic associations and traditional forms of local community engagement, even as online activism has opened new, more transient pathways to civic activity.

The year 2016 brought many shocking developments, but one appreciated by relatively few was the publication of a landmark work of political science by Christopher H. Achen and Larry M. Bartels. It may end up being one of the most important books in at least a generation about democracy and may change the way we think about democratic systems.[43] The authors critique the dreams of democracy with a blast of realism. They present a sweeping survey of empirical and theoretical work over the past century and demolish the "romantic" idea of the rational voter and an electoral system that in any way produces a responsive government. Our civic religion, our "folk theory," about how democracy works or should work is wrong, they argue. No amount of facts or persuasion matter much in how people make up their minds about issues or policy. "Group politics" define people's views on issues and candidates; social-psychological factors and identity-related concerns are dominant, not economic or issue-based ones. Democratic decision-making is, then, a thoroughly social, not rational, process. What this implies for the press is somewhat alarming: the idea of improving democracy by better informing citizens may not be realistic.

The notion that the press is necessary for democracy insofar as it keeps the public informed is central to the mythologies of the news media industry. But, as Achen and Bartels demonstrate with overwhelming data, more evidence and factual analysis is not likely to improve things or facilitate more rational decision-making. The authors offer no programmatic solutions to this core problem, but they do note that making democracy more democratic will require a "greater degree of *economic and social equality*."[44] If democracy will always be riven by group politics and characterized by tribal clustering, the only way to achieve more democracy is to strengthen weaker groups relative to dominant interests.

For journalism, these insights should strengthen a collective resolve to redouble accountability and watchdog investigative projects—thus putting pressure on powerful groups. But they should also prompt us to think about how best to reach unheard voices and groups and connect their concerns to public affairs debates. We should be keener to cover and seriously engage with more grassroots activity, as this is likely much more important over the long run for democracy than political and policy debates among elites. If journalists are to help democracy work, they will need to look closely at the social networks and group politics that provide the basis for most decision-making in our society.

Above all, the challenge of Achen and Bartels's political realism might remind journalists that ultimately the most important role of news media—and what it does best—is directing attention. "We should ... worry less about the press's ability to inject factual information and the public's ability to store it, and worry more about what the press thrusts into the public view and whether this material provokes thought and discussion relevant to public matters," the political scientist Thomas Patterson and journalism scholar Philip Seib assert.[45] This is not to say that facts don't matter, but rather that story selection—and who these stories attempt to connect with—matters enormously. Making good judgments on what to cover is, in effect, most of the ballgame. "Agenda setting" has a pejorative connotation—yet attention guiding, as we might call it, supported by knowledge and smart, well-contextualized reporting can be journalism's most effective function.

In the view of some media scholars, neglecting this reality—that identity and group politics may matter much more than information—has ongoing consequences for journalism in terms of influencing funding, resources, and direction. Daniel Kreiss writes that "journalists, a network of foundation funders, and academics alike generally see the profession of journalism in the narrow and ideal terms of providing quality information to rational, general-interest citizens fulfilling their solemn duty of making informed decisions at the polls."[46] This narrow lens "has placed significant limits on our ability to imagine a way forward for journalism and media in the Trump era."[47]

Tocqueville's Echoes

Given these emerging dangers of fragmentation and disintegration in political and civic life, it is only logical that journalists begin to think more consciously about how their work can help along these lines: how it can play a useful role in civic-oriented network formation, making disparate individuals and groups more aware of one another and their common concerns. News has always been intimately tied to the rise and fall of civic group ties and associations—social networks—as the French political writer Alexis de Tocqueville noted nearly two centuries ago in his journeys across America: "There is ... a necessary connection between associations and newspapers. Newspapers make associations, and associations make newspapers; and if it were true to say that associations must multiply as quickly as conditions become equal, it is equally certain that the number of papers increases in proportion as associations multiply. Thus, of all countries on earth, it is in America that one finds both the most associations and the most newspapers."[48]

I continue to be struck by how profoundly this idea resonates despite all the industrial, technological, and social change that has unfolded since the time of Tocqueville. I attended the fall 2017 meeting of the New York Press Association, the nation's oldest and largest such organization, and presented some findings from a survey I conducted with their membership. (The association was founded in 1853, about two decades after Tocqueville came to America.) Most members are relatively small owner-operators whose weekly papers continue to be both the community bulletin board and the social glue that keep so many towns informed about collective goings-on. Most are struggling with diminished advertising revenue, as one would expect, but many remain determined to stay in operation for one reason: community continuity.

With the evident pride of small business owners who value something more than the proverbial bottom line, second- and third-generation newspaper owners articulated versions of the same mantra: "I won't let it go down on my watch. It's a community pillar." The survey I conducted included the following question: "How much do you believe your news outlet fosters community engagement in civic actions such as voting, town meetings, or participation in events?" Among seventy owners formally surveyed, fifty-four answered "a great deal" or "a lot." The interconnection between associations, participation, and news remains strong, despite the passing of the centuries. And yet, asked to forecast their longevity as a business, only about half of these small newspaper owners thought they would stay in business beyond twenty years, with roughly a third predicting their own publications may fold within a decade.[49] These are publications, in many cases, that have been around for more than a century, anchored in ancient and storied villages across New York state. The news business is in crisis, and its effects radiate far beyond national elections.

For at least a decade, there has been breezy speculation that social media platforms and Internet companies that vacuum up public information might replace many of the information functions of local newspapers. Perhaps this will be true in some part. But one anecdote, relayed over dinner at the New York Press Association, reminded me of a kind of ethical integrity that social media will never replicate: A small town editor, having had too much to drink one night, made a poor decision and decided to drive home from the bar. He was arrested for DUI. A day or so later, he walked ashen-faced into a newsroom of silent employees and, following his code of ethics of impartiality, dutifully typed up his own crime blotter section, writing about his own foolhardiness. It is hard to imagine this kind of civic duty ever being carried out in the history of Facebook. This embedded ethical code and its manifestation in news stories is what the journalist Alex S. Jones has called the "iron core" of news; it embodies an ethical proposition without which the field of journalism becomes lost in the sea of so much public relations and marketing.[50]

Information may want to be free, but it doesn't always want to be ethically guided. Journalism institutions and their norms mean something. And yet we can't afford to be nostalgists. The ethical conduct of reporting and journalism must be attached to new ways of doing things. Technology has radically altered the landscape for journalism. The marketplace and the economics of commercial media have shifted dramatically. How to sustain this "iron core" in the age of networks is in some ways the central question. The way to do this, I would argue, is to imagine how network formation can be guided ethically and knowledgeably by journalists.

2 News, Knowledge, and Civic Virality

News producers are finding themselves increasingly in conversations with networks. The following types of circumstances are not atypical: A web editor for a local TV station posts a teaser on Facebook about an upcoming story. The station is going to run an evening news segment about slow response times to 911 calls, based on the experience of a few individual sources. Within an hour, hundreds of people have commented on the Facebook post, sharing similar experiences. The station ends up doing a five-part series based on ongoing feedback from the public. Authorities are forced to account for response delays. Similarly, in another city, a reporter goes on air to report that someone has been overbilled using the city's parking meter app. The reporter immediately receives dozens of social media and email messages of similar complaints. She then makes public records requests that show a pervasive pattern, generating more stories. The issue takes off like wildfire on social media, forcing public officials to apologize and issue refunds. Many more news segments follow.

This kind of public engagement around such stories was theoretically possible in the predigital era—but it took time to develop. Now, in the era of the social web, these networks of recognition can be generated, accelerating public knowledge very quickly. And speed matters: it can change the fundamental orientation of public knowledge and social networks.

To understand the evolving dynamics of how social networks and news interact, let's consider some deeper case studies of how network formation can take place. We can study the power of networks of recognition—the sudden bursts across the civic domain that create vast social ripples and bring disparate persons into common conversation. As opposed to the kinds of "thin" or superficial Internet virality we associate with cute cat photos, bears climbing trees, and music video parodies, we might consider more deeply the types of *civic virality* that nourish democracy and generate networks of recognition. We might look at patterns that show generativity—of conversation, learning, sharing, and, ultimately, recognition.

To begin, let's look at three different patterns: journalists connecting the crowd, the crowd connecting journalists, and a hybrid pattern in which journalists may more carefully structure their connections to the crowd. In these examples, public knowledge enlarges in proportion to the size of the crowd. We see various forms of recognition, including the grasping of new realities, the validation of others' experience, and the acknowledgment of mutuality.

Spotlighting Networks

Let's start with an example in which journalists help connect the crowd. In 2002, the *Boston Globe* exposed a systematic cover up by the Catholic Church of priests abusing children, touching off a global reckoning for the church laity and leadership alike—and resulting in the award of a Pulitzer Prize for the newspaper. The story of how the *Globe*'s investigative unit unraveled this story of secretive networks has now been made famous through the Academy Award–winning movie *Spotlight* (2015). But there are still a few big mysteries. How had a conspiracy of these proportions, of this magnitude, stretching across the globe to hundreds of parishes, remained hidden for so long? And why had a story first published in January 2002 in a mostly regional publication, the *Globe*, so quickly touched off a maelstrom that began consuming church communities globally within a few short months? Why did it take so long?

It is useful to visualize the pattern of information flows during that consequential year of revelations (see figure 2.1).[1] In retrospect, it represents a watershed cultural moment in terms of opening a greater dialogue about the many forms of abuse endured by disempowered groups. It created a kind of network of recognition that stands as a template. It is therefore important that we try to understand how this knowledge spread and how the culture began to change in relation to news production.

As those who have seen the film *Spotlight* may recall, the breakthrough insight in the investigation came when the reporting team—Walter Robinson, Matt Carroll, Michael Rezendes, and Sacha Pfeiffer—realized that the network map of the priests in question was discernible in the annual

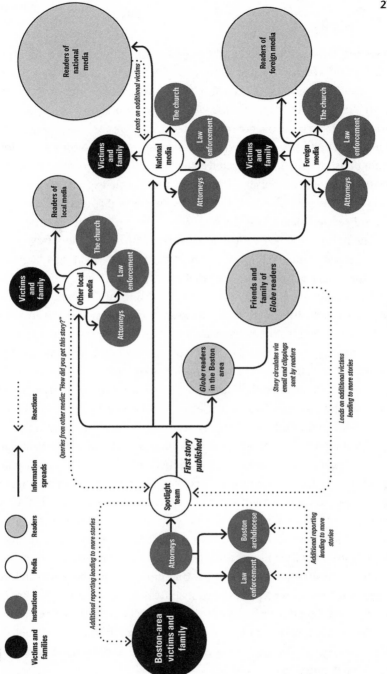

Figure 2.1
Networked information flows during the 2002 *Boston Globe* investigation of the priest abuse scandal.
Source: D. Zedek, M. Carroll, and J. Wihbey.

directory publications produced by the Boston archdiocese. By finding all the instances of priests listed as going on "sick leave," or other coded terms, the reporters could produce a list of likely suspects. The reporters painstakingly compiled the various case data points into spreadsheets, counting and organizing all of the potential instances of abuse across the greater Boston area. They retraced the networks of abusive priests as they were moved from parish to parish by church leaders. All of this inquiry also was spurred by the reporters' conversations with knowledgeable experts, who had long suspected the pervasiveness of the problems (even if a full data picture had never been rendered to the public).

The priest abuse scandal was a network problem—in many ways, a classic type that would be familiar to crime fighters investigating a large illicit enterprise, performing what is sometimes called *link analysis*. A vast web of priests and victims was connected by invisible ties through the network of the Church. Information about these cases was aggregated in certain "hubs"—officials in parishes and archdioceses, the secretive warrens of Vatican officials—but the nodes (points representing persons in a network diagram) at the margins of the vast network were disconnected. These silent nodes were victims and their families. The network had been used to cover up abuse by actively discouraging any connection among isolated cases of abuse. Accused priests were quietly moved to other parishes, and victims quietly coerced into out-of-court settlements, with nondisclosure agreements attached. Links were frozen in an inert state; nodes remained isolated.

As the number of potential priests involved in wrongdoing began rising, from three to eighty-seven, the reporters knew they had a giant story. But until the first story ran, there was no way to realize quite how big a story it was. Indeed, what they had discovered was not just a ground-shaking Boston story but a small piece of a latent network of many dimensions that extended around the world. Why, at that moment, did the story catch fire? There had been other reports of priest abuse in the press, after all. The *Globe* stories would produce what is one of the first instances, characteristic now of the digital age, of a true news-driven "information cascade" fueled by a peer-to-peer technology—namely, email, one of the elements of what's sometimes called "dark social."[2]

The initial stories and links to them online were forwarded many times over email. A decade earlier, no such phenomenon would have taken place. After the first story ran in 2002, thousands of people who were forwarded the story then emailed the *Globe* reporters with their own stories. The emails came from across the country and around the world. After the first story ran, reporters had victims contacting them the very next day from as far away as Australia. This is simply unimaginable in any prior era of information and communication technology. By then sharing these stories, the *Globe* indirectly helped build the offline survivor groups that would prove key to legal and church reform. Reporters from around the country, and around the world, also began emailing the Spotlight team, asking for help in breaking similar stories in their own geographic beats. A template had been established and a code broken. A new symbolic environment had been created that facilitated the creation of an interested public, empowered through network formation. The new symbols and framings of victimhood and empowerment, cover up and justice, generated an iterative process of new revelations and new meanings for an expanding number of communities.

News media can, at their best, provide an attention platform, a form of social power and political space that can counter other power structures. They also can provide a new language and framework, which, when supported and embedded in knowledge, can shift societal narratives and understandings. This can strengthen networks and make information flow more freely. For abuse victims, this new language meant news media helping to publicly reframe their feelings of shame and guilt as the consequences of moral wrongs perpetrated systematically against them and recasting these experiences clearly as instances of intentional harm, illegal acts demanding prosecution. The key was showing that the crime wasn't just the product of a bad priest here and there. It wasn't just an anecdote, and it was no longer a hushed rumor, easily dismissed. Being able to publish internal communications from the church, as the *Globe* reporters and editors did once they gained access to formerly sealed court documents, was key to solidifying public acceptance of this new knowledge. People could *see* the documents for themselves.

A major journalistic series—and a study in network power—had begun in a prominent place, visible to authorities and the public alike. The latent network was activated, and a permission slip had been given to the various nodes to connect, share information, and form their own counter-Church survivor groups. These permission slips for dialogue and connection are what communication and network theorist Manuel Castells calls "sharing protocols."[3] These protocols govern what can and cannot be shared.

The Spotlight team's leaders, team editor Walter Robinson and editor in chief Marty Baron, knew that getting beyond anecdote and into network hubs, the church power structures, was the only way for the story to cut through. They frequently said as much to their team. Journalism at its best has always felt intuitively the importance of networks of recognition. For generations, reporters have spoken of stories "resonating," of producing "talkers."

The story's evidence was overwhelming and narrative-shifting. The ringing lead to the first installment in the Pulitzer Prize–winning series, published January 6, 2002, was this: "Since the mid-1990s, more than 130 people have come forward with horrific childhood tales about how former priest John J. Geoghan allegedly fondled or raped them during a three-decade spree through a half-dozen Greater Boston parishes."[4] The *Globe* had established systematic information, deep patterns. This is news gathering at its best and most powerful; it had harnessed the stories of the crowd, of dozens of average citizens, and connected nodes together in a way that could overcome countervailing forces of cover up and enforced silence. New data and documents had produced new knowledge in the culture. It was enough to signal to the vast latent network of other persons, most of whom were previously silent, that the power dynamic had shifted. A powerful network of recognition had been activated. The technology of email then facilitated the reverse activation of the exact network that had been used to perpetrate acts and suppress knowledge of them. The network was retraced and rewired to fight back.

The remarkable thing about the *Globe* investigative story of priest abuse and the conflagration it ignited is that it all took place before the advent of what we think of as the true social web era, or Web 2.0. One wonders what might happen now and how much more quickly the story would spread. We might have some indication from the 2017 #MeToo campaign, in which women across the Internet shared their stories of sexual harassment and abuse by the thousands, seemingly instantly.

A combination of technology and connected newsgathering and storytelling can be consequential in stirring citizens and firing grassroots activism. There are important, mutually reinforcing dynamics among the moving pieces: the network-oriented reportorial hunt for sources; the heightened public awareness and empowerment through the process; and information and communication technologies that accelerate the related feedback loops, creating a kind of "compound interest" of knowledge about issues. Because of these mobile-enabled, social technologies, the flow of information in society is changing all the more. Open, visible social platforms now allow public storytelling by persons and groups of all kinds, shifting the power dynamics of media. Fast-forward a dozen years beyond the priest abuse scandal story and we see new such dynamics at work.

Authority of Counterpublics

Now let's take a look at an example of the crowd triggering journalists: the "hijacking" or takeover of the #myNYPD hashtag in 2014 by activists and citizens who were protesting police-involved shootings and other forms of aggression against minority populations. Between April 22 and 24, what began as a feel-good public relations campaign for the New York City Police Department, which encouraged citizens to share pictures of themselves with police officers, evolved into a poignant, symbolic moment in which citizens and activists registered a wide range of complaints about law enforcement. People used the Twitter hashtag to show images of citizens being arrested with questionable amounts of physical force and images of persons being injured or killed by police. The powerful symbolic shift in meaning of that hashtag quickly cascaded around the world, generating other copycat memes—from #myLAPD in Los Angeles to #MiPolicíaMexicana in Mexico to #myELAS in Greece, and far beyond.[5]

The case of the #myNYPD takeover provides a window into the paradigm shifts in media power we are now increasingly seeing. The information sources at the center of the network were mostly individuals and collectives of activists, not news outlets (see figure 2.2). My Northeastern University colleagues Sarah J. Jackson, a communication scholar, and Brooke Foucault Welles, a network scientist, have studied the patterns of networked communication during that three-day period, concluding that it shows how persons with relatively little broadcast power can, under certain conditions, get attention from the public in dramatic ways.[6] In the network map they draw, one can see individuals' Twitter accounts —@KimaniFilm, @MoreandAgain, @MollyCrabapple—and activist collectives—such as @CopWatch, @OccBayStreet, @OccupyWallStreetNYC, and @YourAnonNews—with central positions, based on the number of mentions and retweets. The map looks like a broadcast network—except that the central "hub" positions are not mainstream news sources, but rather individuals and grassroots groups. The mainstream accounts of @NBCNewYork, @BuzzFeed, @Vice, and @ AJStream (Al Jazeera) are there in the diagram, but the central power in the network is not mainstream media. The true source of the new knowledge, the new symbolic reality that enabled new forms of communication to unfold, was a set of grassroots actors. Eventually, mainstream news outlets caught on to the story and helped propel news of the protest across the world. But journalists had not been sufficiently attentive to these still-burning sentiments across New York, a city that has seen decades of controversy over police conduct.

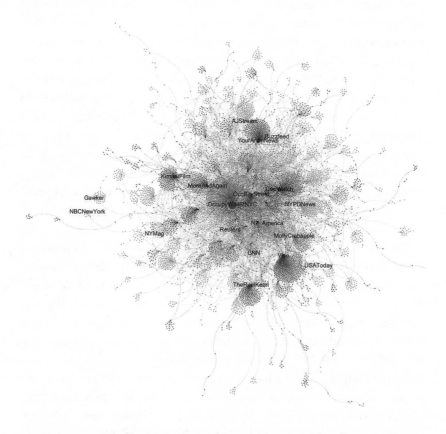

Figure 2.2
Hijacking/takeover of a hashtag, #myNYPD, by activists highlighting police brutality in New York City (2014).
Source: Sarah J. Jackson and Brooke Foucault Welles.

This flipping of traditional media power dynamics fits in with a broader theory of how *networked counterpublics*, marginalized groups who do not typically have influence in the mainstream public sphere, may exert more power by using online tools to challenge "dominant knowledge." In this case, citizens and activists were challenging and unsettling the very understanding of the NYPD and its operations and impact and what it all meant to the communities in the city. "That diverse individuals with relatively few followers," Jackson and Foucault Welles note, "were elevated to a similar level of visibility in the hijacking of #myNYPD as organizations with far more followers supports the existence of democratic access in networked counterpublics."[7] New perspectives and forms of knowledge—conveyed through vivid and disturbing pictures, as well as first-hand accounts—created a new issue space that generated a cascade of novel information. The connectivity of a network of diverse and disparate persons increased rapidly.

What such incidents suggest is that journalists should be increasingly attuned to the information needs of citizens as expressed through online platforms. Social media provide a powerful new system to signal what is important to communities. Certainly, not everyone in New York City felt that it was a fair representation of police work to feature scores of unflattering images, as the activists did, and journalists must always be sure to verify crowdsourced information and put it in context. Indeed, journalists' primary value in any controversy is the sort of careful, knowledge-based work that should be expected of them as professionals. But nevertheless, the physics and geometry of news are changing. The #myNYPD takeover event helped begin to change the symbolic reality for the public around such issues, bringing increased connectivity to social networks on the topic of police brutality and police-involved killings. A nascent network of recognition was forming. Just a few months, later major incidents involving police and the deaths of young black men would take place in both Ferguson and Staten Island that would touch off a national conversation that is still reverberating.[8]

If news media are to maximize their usefulness and purpose in society, they must take seriously the new dynamics of networked knowledge and harness them to help improve democracy. Agenda setting in media is changing in important ways; as mentioned, it is now more about attention-guiding. "Thus the new journalist is no longer deciding what the public should know—the classic role of gatekeeper—but working with audiences

and technology to make order out of it, make it useful, take action on it,"
Bill Kovach and Tom Rosenstiel note in *The Elements of Journalism*.[9] Their
formulation—and Rosenstiel's related idea of "news as organized collaborative intelligence"—usefully highlights this evolving role for media as cocreators of networks of recognition.[10]

Health Links

Consider one last case study in civic virality, one in which journalists build
ties across a far-flung, disparate network that, unbeknownst to many individual nodes, has experiences and problems in common. It shows journalists building upon and humanizing systematic knowledge to establish the
priority of an issue and helping a nascent public to form mutual awareness
and understanding. Here we see journalists leveraging the crowd to trigger the crowd, which then leads to a compounding, growing dynamic of
enlarged knowledge.

The story began with awareness of a troubling trend in maternal outcomes during childbirth in America, but it evolved into a story of networks,
of disparate human stories being gathered and synthesized together over
time. When journalist Adriana Gallardo of the investigative news nonprofit
ProPublica began working on the story behind America's high rate of mortality among mothers giving birth relative to mothers in other rich countries, she and her team were tasked with formulating a strategy for finding
individual stories that helped to surface some pattern of explanation for
this tragic phenomenon. State-level information on maternal mortality in
the United States was sparse; there were few public updates on who was
dying, where the cases were located geographically, and specific causes of
death. The ProPublica team, working in partnership with National Public
Radio (NPR), became aware of the statistical patterns and the latest peer-
reviewed research that raised myriad questions, but they could make little
sense of what the real medical and human story was behind the numbers.[11]

The journalists could see there was a definite "information gap"—a fundamental hole in society's knowledge and the human network surrounding
the issue. Gallardo, whose title is *engagement reporter*, works at the intersection of community- and relationship-building and gathering of information
for reporting. She and the reporting team began with a two-pronged strategy: first, a "structured call" for people to contribute their stories through

an online form and an email address; and second, massive scouring of the Internet, using techniques of "social verification and social research," to try to track down stories of women who had died in childbirth. They were using Facebook Signal (an algorithmic news-finding tool) and Twitter, as well as looking at crowdsourced sites such as GoFundMe, which pointed to promising leads.

Before the first eight-thousand-word exposé, or any of the other incredible reports in the series, was ever published (the first story came out in May 2017), the story began as a social research project, in collaboration with lead reporter Nina Martin, who had extensive experience reporting on women's health issues.[12] The reporters worked with experts to develop their online survey, carefully crafting the questions and fine-tuning the information-gathering strategy. More than 2,500 women and families responded in the first week to their online appeal for information, sharing stories and pictures. A video shared by a grieving husband, used with permission by ProPublica, was repackaged as a video circulated on social media and viewed more than three million times in the first weeks after initial publication.

Deaths of mothers were being reported within the online network that ProPublica was hosting even before they even showed up in obituaries. "The way that we're working now is that we take each investigation on its own, and think about what ways we can include community, and non-traditional experts, in the reporting process," Gallardo notes. "We're doing these very tailored engagement efforts that work in tandem with the timeline of the story."[13] The maternal mortality story has sparked all sorts of discussions among families and individuals who have had common experiences and led to community discussions with doctors and hospitals. Spurred by the stories, mothers who have had near-death experiences have found each other on social media.

ProPublica has created Facebook groups related to investigations such as this in the past: for veterans' issues, for example, relating to the use of Agent Orange in Vietnam and its ongoing effects on service members and families.[14] The dataset of affected families relating to childbirth incidents is also being used by ProPublica to engage individuals and families with related issues of health care policy and race, for instance.

An important distinction between this approach and traditional approaches to reporting involves the element of time and sustained attention. "The difference is that we're not leaving throughout the story process," says Gallardo. The maternal mortality series in effect "embeds" sources in the story process. The network of recognition that is formed, with journalism as the hub and connector, is sustained over time through a series of stories and public conversations. And the metric for success is not just aggregate pageviews, she says, but rather, "Are we building the right communities around each investigation?" It's an essential question for the future of journalism practice.

The ProPublica story might be seen as a textbook case of what is broadly being called the *field of journalistic audience engagement*, or *engaged journalism*, in action. It speaks to the emergence of a new reality and opportunity as digital platforms have enabled journalists to find relevant sources and audiences more quickly and at scale. The techniques and philosophy employed in the ProPublica story and series join a debate over how much news should increasingly be seen as purely a kind of "conversation" rather than as a product being delivered.

We are now about a decade into this new phase of user participation, the Web 2.0 revolution exemplified by Facebook, Twitter, Reddit, and the like. Some reporters have made a distinction between the practice of mass crowdsourcing and more "organic" and targeted uses of social media to develop information and sources.[15] Media researchers have been cataloging the various typologies of engagement, which include everything from "solutions-focused collaboration" to "public convenings" with community members.[16] Best practices are emerging, even as the capacity of technology opens up ever-more new possibilities and experiments.[17]

Although many news outlets have not embraced this model yet, it is the theory of change embedded in innovative new research efforts to define and encourage audience engagement in journalism. It guides the work of the Coral Project, Hearken, GroundSource, the University of Oregon's Gather project, and Democracy Fund's Public Square Program.[18] This new model has also helped fuel the rise of the position of engagement reporter or editor. Contemporary journalism circles have focused increasingly on the concept of *engaged journalism*, an evolving set of theories and practices that specify direct interaction with news audiences throughout the lifecycle of stories.

With these practices, some news practitioners are squarely addressing some of the core problems of democracy: the lack of a public sphere with civil debate. Public radio station KUOW in Seattle, for example, has created in-person "speed dating" events, in which audience members meet each other to discuss issues and deliberate.[19] Employing a new model called *dialogue journalism*, the Spaceship Media project is trying to facilitate dialogue among thousands of people—across diverse topics, from agriculture to education to immigration—by moderating large-scale social media groups. The project's goal is to foster deeper conversation on polarizing topics.[20] Outlets are also using engagement techniques to address the public's lack of trust in news media. Cincinnati TV station WCPO, with guidance from the Trusting News project, is being more open with audiences about how it is putting together stories and more engaged with the community that responds to and discusses the stories.[21] Media reform groups are even trying to create engagement infrastructure and solutions at greater scale. The News Voices initiative in New Jersey, a project of the nonprofit advocacy group Free Press, is trying to connect local citizens directly to newsrooms across the state, building community connections by surfacing "stories that should be told."[22]

Varieties of Knowledge and Recognition

Let's now take a step back from the world of media practice to understand a bit more about the fundamental dynamics of networked information flows. Knowledge and networks interact dynamically and often unpredictably, producing emergent phenomena. Networks of recognition produce bursts of knowledge and awareness. The flow of information affects the volume and spread of what is known; new knowledge is produced as networks of recognition expand.

The very terms *knowledge* and *networks* are semantically porous, and both are thrown around so frequently in discourse about the twenty-first century world—think of *knowledge economy*, *age of networks*, and so on—that it is worth clarifying these terms. There are several layers of meaning that are relevant to our discussion. These are worth distinguishing carefully, as their definitions are different depending on whether we are referring to the journalists or the public. We need to distinguish between scientific knowledge and the social epistemology that I have mentioned.

First, I am arguing that *knowledge* in the traditional, scientific sense has a very strong—indeed, vital—role to play in improving professional practice. Knowledge in the context of journalism, in the construction of news, might be defined as information that reflects some representative pattern in the world, or "systematic information."[23] Methodologically, it is different than information derived from a reporter's "person on the street" interview or from a reporter involved in the act of bearing witness to a particular event. At the level of journalism practice, it may involve the practice of data journalism or the drawing on bodies of knowledge such as peer-reviewed research to inform a story.

In this sense, knowledge is the vital context into which particular human stories can be fit and understood. In the priest abuse scandal, it was tapping experts who had estimates of the scope of the problem and then proving such hypotheses through data collection. For issues of police-involved killings, it was finally understanding the prevalence of the problem and broad community perspectives on its effects. With respect to maternal mortality, it was reviewing the research and statistical picture while also filling in gaps in understanding by soliciting large volumes of geographically dispersed stories. With more systematic context, anecdotes then make sense. For a journalist, knowledge is the puzzle itself or the mosaic. The anecdotal story is the piece or the tile placed within the puzzle.

For journalists, knowledge is an instrument in fostering recognition; it is the vital fuel and the context for recognition. In democratic cultures, recognition also has a deeper, political-ethical dimension, as Allen observes: "Political friendship begins from [a] recognition about what we share with the people who live around us and in the same polity. It moves from this recognition of a shared horizon of experience not to a blind trust in one's fellow citizens but rather to a second recognition that a core citizenly responsibility is to prove oneself trustworthy to fellow citizens so that we are better able to ensure that we all breathe healthy air."[24]

This "shared horizon of experience" is the basis for network formation. Recognition is the fuel. Some journalistic work may not initially foster networks of recognition, of course. Breaking news stories, for example, may not provide an immediate platform. But after the waters have receded, or the immediate crisis of whatever kind shifts into a new phase, journalists must work to show how many different levels of society are implicated, knit together in core citizenly responsibilities and decisions.

An interconnected web is involved in all complex issues. In the priest abuse scandal, there was a network that ran from politicians, the courts, and the Vatican down to families and the Sunday school teacher. In the context of police-involved shootings, a conversation needed to unfold at many levels, across many policy domains: race and justice, social services, mental health, gun policy, training of authorities. The ongoing story of maternal health in the United States likewise requires recognition of the varying roles and responsibilities at all levels of the system. The highest good—trust—will not always immediately flow, but recognition is a first step toward trust.

However, in terms of understanding the public and its changing relationship to news, knowledge also has an important secondary meaning in this discussion. Here we return to social epistemology. Knowledge pertains to shared information, items that are "known" or recognized within a community. As mentioned, Schudson has noted that news itself is a form of what might be called "public knowledge," which serves to construct a "symbolic world" whereby items are given priority or importance to publics.

Journalists in the digital age are increasingly both producing very short snippets of news and conveying fleeting impressions (e.g., tweets or news alerts) while also producing incredibly in-depth, research-based work.[25] In this regard, the philosopher William James made a useful distinction between "knowledge about" (more formal) and "acquaintance with" (more informal.) "Knowledge about a thing is knowledge of its relations," James noted. "Acquaintance with it is limitation to the bare impression which it makes."[26]

Both kinds of knowledge, occupying a spectrum from highly formal to the briefest of impressions, are relevant to the discussion of social networks and how they are activated. The idea of recognition can encompass this spectrum of knowing, from knowledge about to acquaintance with.

As will be discussed in more detail in chapter 3, the age of digital, decentralized networks changes the basic calculus for public knowledge acquisition and the encounter with knowledge. Digital networks change what Hardin calls the "economics of ordinary knowledge."[27] The web allows news to find consumers, and it allows consumers to access knowledge instantly. This has implications for the role and purpose of news. "By an economic theory," Hardin notes, "I mean merely a theory that focuses on the costs and benefits of having and coming to have knowledge, or to correct what knowledge one has. It must fundamentally be a theory of trade-offs between gaining any kind of knowledge and doing other things, such as living well."[28] We might see vast gaps in understanding among the

public—think of the proverbial inability to name one's elected representa-
tives, or to vastly overestimate the amount of US foreign aid—as rational in
the sense that members of the public are seeing knowledge acquisition and
the time it takes to know a given thing rather pragmatically.

We now have two working definitions relevant in our discussion—knowledge as systematic information, as scientific in character; and social epistemology or public knowledge, both knowledge about and acquaintance with.
It should be said that there are always tensions and complications between
these definitions of knowledge in the practice of professional journalism.
This is because journalists are constantly dealing with anecdotal information and (one hopes) trying to reconcile it with systematic information.

A journalist who makes an appeal on Facebook for information from a
digital crowd or who reports by knocking on doors in a neighborhood may
get many powerful anecdotes from groups. Perhaps tensions with police are
revealed or stories of inadequate public services. The journalist may then,
in effect, knit together those persons and their anecdotes, potentially forming a new, or denser, network of people and information points. However,
those anecdotes may not amount to any representative pattern; they may
be outliers, even if the stories are themselves valid and important. What
we might count as knowledge in the context of network formation may
not rise to the level of social scientific information or anything systematic.
This is when, as will be discussed in detail in subsequent chapters, journalists must exercise careful judgment, acknowledging uncertainties and being
transparent about the extent or lack of evidence.

Journalists must successfully navigate both the crowdsourced and the
systematic worlds: to interpret accurately, to choose what is important from
the crowd and properly contextualize it, they must draw on knowledge in
the systematic and representative sense. We might distill this to a single
idea: *to choose, one has to know.* Consider the situation now faced by journalists at every level, from the local correspondent to the global editor.
The problem is no longer scarcity of information but abundance. Once the
journalist might have found her primary work focus in seeking to interview
several people, but now a thousand voices may already have weighed in on
the relevant issue on social media. This puts a great burden on news judgment, on journalists' capacity to select what should be prioritized as public
knowledge. This reality demands a well-grounded grasp of issues, of systematic knowledge, particularly given the deadlines and pace of news and

information. The primacy of selection demands more such formal context and knowledge if journalists are to improve the quality of their practice, avoid the mistakes of the past, and work to foster networks of recognition.

As a culture, we are transitioning from a notion of news, of mediated pictures of reality, as fundamentally something that *informs* to something that *recognizes*. The role of media is increasingly to facilitate recognition. Recognition can be a universal good, but it is also infinite in its potential objects. We require some criteria to distinguish between objects in urgent need of recognition and those that are not urgent; we also need to be able to distinguish between legitimate and illegitimate struggles for recognition. Media guide the provision of attention. This capacity is inextricably linked to the notion of recognition. To organize the flow and creation of recognition, knowledge must be applied in decision-making about the distribution of attention.

Although formal knowledge is necessary for intelligent selection of information by journalists, the task of connecting publics on issues of shared concern involves journalists being able to make relevant information known to communities of interest. On digital networks, the mutual "acquaintance with" translates into a kind of ambient readiness for further dialogue, for recognition, and for the spreading of related information among members of the public. Through this act of shared knowing, groups or networks are knitted together through recognition, through more or less strong or weak ties. Latent ties are activated; new ties are created. Knowledge in this sense is a necessary condition for group discussion and action.[29]

Defining Network Patterns

The word *network*, as we will discuss in chapter 5, is an old one, dating back in English to the 1500s. It has a rich linguistic history, despite the novel feel of its usage in the digital age. It was during the era of the early telegraph that the American public, marveling at the dense tangle of wires strung overhead across the nation, began to use the term *net-work*, originally hyphenated, in the communications context.[30] The most important meaning for our purposes relates to any social network that is linked together or produced by information and communication technologies, whether Facebook, Twitter, email, or any other electronic platform that allows for peer-to-peer connectivity.

Networks can be what scientists call *multimodal*, insofar as they can have different kinds of nodes, such as editors and Wikipedia articles, or people and institutions. Networks can also be *multiplex* and have different kinds of edges, or connecting links that represent different purposes, such as friendship, cooperation, or expertise. The study of networks also can involve graphing the relations between objects (sometimes called *link analysis*), with objects of various kinds mapped in relation to one another. Counterterrorism and criminal investigations often construct such networks to explore the relationships among individuals, communications information, locations, bank accounts, and other data points.

We can also imagine graphing how points of knowledge and information are connected to other knowledge and information nodes. These correlated knowledge points might be highly structured. For example, academic papers might be mapped around a given topic; they also might be brought together with commentaries and other kinds of articles. They might be contained in a relational database that forms the basis of a news app (e.g., a map of properties with environmental contamination in a county) or a Wikipedia article that is built on published sources from around the web and curated by platform users. Or perhaps the nodes are just "loosely joined," as the technologist and philosopher David Weinberger puts it, through scattered hyperlinks or Twitter hashtags.[31] There are hybrid possibilities, too, for networks in which nodes might represent both persons and knowledge points.

Quality networked journalism generates new publics, creating bridging ties between disparate communities and forming links across what network researchers sometimes call *structural holes*: relational and knowledge gaps between groups and clusters of individuals. Journalism should understand the broader framework around this pattern. The social science theory in this area has been set by foundational figures such as Elihu Katz, Paul Lazarsfeld, and Robert Merton, who observed how ideas can flow through connecting opinion leaders, as well as Stanley Milgram, and his idea of surprising "small-world" bridges between disparate persons, and Mark Granovetter, who famously noted the power of "weak ties," or persons who are only casual acquaintances, for searching for job leads.[32]

The original theorist of the structural hole concept, Ronald Burt, noted that bridging such gaps can provide a competitive advantage within organizations. The tendency for humans to cluster in narrow social groups is well established in the social sciences, across many types of activities. "People focus on activities inside their own group, which creates holes in the information flow between groups, or more simply, structural holes," Burt notes. Further, the "simplest act of brokerage is to make people on both sides of a structural hole aware of interests and difficulties in the other group."[33] Bridging in this way generates social capital. Beyond mere links to create awareness, there are increasingly higher levels of this form of arbitrage between disconnected groups: transferring best practices; drawing analogies between ostensibly dissimilar groups; and, at the highest level, synthesis that involves combining ideas from groups. Burt's ideas, devised in the context and domain of organizational behavior and strategy, can be applied and appropriated more broadly as one of brokerage among publics, tying together disparate groups for the purposes of forming closer and richer ties between them. The social capital that is generated can in this way be shared.

Not all of these capacities are entirely new, but digital networks make them possible more frequently. In decades past, journalists often sought to bring certain societal problems to the attention of authorities, bridging the structural hole between citizen grievance and policymaker. Journalists are connection points that, in important ways, broker knowledge and act as cluster-attracting hubs to which many disparate persons, or nodes, attach themselves through media pathways, links, and streams. Sometimes this involves a journalist bearing witness to events and relaying them to the public; other times it involves digging out arcane knowledge, in government documents or little-known corners of the corporate world, and connecting that knowledge to relevant publics, both citizen and policymaker. It now involves journalists enlisting large numbers of people on social platforms to share information and form communities of knowledge, bridging structural holes.

This precise dynamic in the context of journalism has begun to be acknowledged by journalism scholars. "From a network analytic perspective, journalists occupy a bridging position between two worlds that are structurally unconnected: the sources of news and the members of the audience," communication and media scholars Pablo J. Boczkowski and Eugenia Mitchelstein note. "Reporters in particular, and news media in general, act as bridges in that they usually are the only path between those two clusters for most audience members."[34]

Further, such observations are being coupled with related insights for media practice. Journalism scholar Sue Robinson notes that it is the "job of the professional communicator today to learn the best ways to manipulate social networks so as to build connecting and connected bridges instead of exacerbating rampant distrust that nurtures isolation and polaritization."[35] This new mission of connection has a still-developing science behind it on which journalists might draw.

Activating Phase Transitions

Behind the seeming randomness of our age of networked communications and viral media lie regular patterns and laws. These help explain how information swiftly cascades and large audiences gather. When attention suddenly shifts and communities form rapidly, what we are seeing is the process of isolated nodes, somewhat powerless or disengaged in their relatively disconnected state, transitioning from a phase of disorder to a kind of rough order. The new links stitch together a community of mutual interest, attention, and concern. In networks, structure is vitally important— often far more important than the inherent attractiveness of the given informational content, in terms of explaining why a given idea or story "caught on."[36]

The network scientists in natural sciences domains such as chemistry and physics talk about a *phase transition* taking place, a universal characteristic of networks. Such transitions see a network that is disconnected suddenly, with a few well-placed links, become entirely connected, creating what is sometimes called a *giant component* (a massive connected entity, as opposed to multiple disparate groupings). The point at which this transition begins to occur is called a *critical point*, and there are mathematical laws that govern the nature of such network changes from disorder to order.[37] The molecules of liquid water, for example, very rapidly transition at the critical point and create ice. In social networks, such a rapid increase in the connectivity of a network is necessary for contagion and virality, allowing information cascades to take place (see figure 2.3). This phenomenon helps account for the explosive, or "burst-like," nature of discourse on social media.[38] A phase transition is nonlinear. It appears to happen all at once, but in fact there are underlying processes that were building up to enable that transition to happen with a very small change (e.g., adding a few ties suddenly brings disconnected *dyads*, or connected pairs, into a whole network).

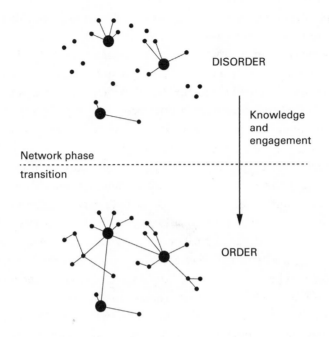

Figure 2.3
Journalism and the fostering of increased connectivity in networks.

Complex societal problems, which are often plagued by public inatten-
tion and policy inaction, are always in some sense network problems. The
network formation necessary to address those problems requires a common language, or some shared system of symbols. "Our Babel is not one of
tongues but of the signs and symbols without which shared experience is
impossible," the philosopher Dewey notes.[39] In the world of social media,
the most obvious symbol has become the hashtag. To adopt a hashtag is,
at its root, to name a problem or concern or interest. To solve a problem,
naming the variables is the key step. It is true in fundamental science: as
it is in mathematics, so too in human affairs. The hashtag names the variable in question. The network forms around it, and through it, and is the
fuel for its activation. Of course, to say that journalists are bringing "order"
to networks can be misleading; they are helping to connect and sustain
memes, themes, issues, and topics. Social networks are dynamic and always
subject to change and decay. Journalists can create the attention structures

to sustain democratic dialogue. Networks of recognition draw their power in part from the burst-like phenomenon of phase transitions.

But what is the key to understanding meaningful bursts of public knowledge, as opposed to mere fluffy viral phenomena? The media scholar James Carey has made a useful distinction that can facilitate greater understanding in this context. Carey distinguished between a "transmission" view of communication—the more common and mundane idea of sending and transmitting messages among persons—and a "ritual" view. The latter is something much deeper and in ways antecedent to transmission. Carey observed that the ritual dimension of communication is "directed not toward the extension of messages in space but toward the maintenance of society in time; not the act of imparting information but the representation of shared beliefs. If the archetypal case of communication under a transmission view is the extension of messages across geography for the purpose of control, the archetypal case under a ritual view is the sacred ceremony that draws persons together in fellowship and commonality."[40]

The ritual model of communication focuses on the construction of shared beliefs and values through the creation of a set of common symbols. Think of the antipolice brutality campaign that began around #BlackLivesMatter or #BLM, or the anti-sexual-harassment/assault campaign using the #MeToo hashtag. The religious overtones of "ritual" may be a bit misleading, in that they imply strict or blind adherence. But this model of communication implies a coming together to create meaning in an iterative process. Gathered together and empowered by new symbols, communities work out and generate new meanings within new issue and belief spaces. This dynamic was noted long ago. Dewey saw the power of the common person engaged in dialogue with community members and believed that the role of the press, if oriented correctly, could facilitate this almost spiritually nourishing form of democratic communication—of "enriching communion" with others.[41]

Digital media scholar Limor Shifman has used the transmission-ritual idea in communications to help distinguish mere "viral" content from more meaningful "memes," in which a given meme's message "is not a unit whose reach and effect are easily traceable, but an ongoing process in which identities and sense of belonging are continually constructed."[42] This process is stimulated by the addition of new knowledge, new facts, and perspectives that are grasped by publics. As the journalism community has begun to recognize the limits of the "transmission" model of communication, and the desire to merely "go viral" and chase online clicks, its practitioners have begun to talk about the central mission of stories as engaging the "right" community, the affected community, or the community of interest. Put another way, this is the potential network of recognition. Its activation requires common symbols.

In this context, consider again the case of police-involved shootings in the United States, an ongoing area of pressing policy concern that was ultimately crystallized by the death of Michael Brown in Ferguson, Missouri. Despite all the gains in racial equality in post–civil rights movement America, problems between minority communities and law enforcement had persisted in pockets around the country, with virtually no national attention and little policy action. In part, this was because the pattern of problems remained diffuse and disconnected—a structural hole. As local activists and then news media began raising awareness of these problems, suddenly a large network of such previously invisible problems was connected, fueling the Black Lives Matter movement and numerous congressional and state-level hearings and proposals. What is remarkable in this case is that, statistically speaking, there was really no change in the number of such disturbing incidents across the United States compared to prior years. What was different was that a network had formed, activated in large part by citizen-produced media images and videos and magnified by news institutions.

As mentioned, a similar dynamic was at work with respect to local water quality problems that in recent years have become salient to the American public, exemplified by the case of Flint, Michigan. Local complaints were being registered frequently, and some were reported in the local press. But no network that could prompt action was formed until it was far too late.[43] Again, numerous communities were facing a range of pollution issues. Once the salience of the Flint case was raised and more journalism in place, it became apparent to the public and policymakers that such problems were pervasive.[44] A hole had been bridged through shared knowledge, and a phase transition had taken place. A network of recognition had been created.

Journalism and Early Detection

In both the police-involved shootings and Flint cases, journalists were inattentive early on, even as citizens took to social media to express their grievances and articulate problems. In listening better to networked communities online, and in looking more intentionally for opportunities to connect disparate nodes in information and social networks, journalists now have enormous power—with far better insight than was possible in the predigital era—to facilitate networks of recognition that effect meaningful change in public knowledge.

Herbert J. Gans has explored how his foundational concept of *multiperspectival news*—a news media more responsive to average citizens and persons outside of established power structures—might be operationalized in the era of the Internet. His original theory, articulated in 1979, called for less Washington-centric news, more bottom-up news, stories on "outputs" of policies and programs, more "representative" news on the activities of the diverse citizenry, and more "service" news on "personally relevant" stories for citizens about the activities of government agencies.[45] As Gans concedes, implementing multiperspectivalism may be no easier now in the digital age than it was in the broadcast age. Yet better tools for listening and detecting important new perspectives are now in the hands of journalists.

As will be discussed in chapter 4, understanding the science of networks can facilitate the interpretation of patterns found in online networks, helping journalists to maintain a healthy skepticism and more carefully discern how messages are disseminated (and suppressed.) Network science is producing a wide variety of insights useful to those producing media that incorporate engagement with and information gathering on socially mediated digital platforms. For example, the study of the concept of homophily (first developed in an offline context) suggests that basic patterns characterize how like-minded individuals tend to cluster online as well.[46] Research on what is called *modular structure* yields insights into how community structures and hidden common bonds might be discovered.[47] Further, the study of algorithmic filtering can help alert journalists to how computational systems may skew or bias how citizens understand reality and how the flow of ideas may be affected. Algorithms certainly had a role to play in the lack of visibility of important issues in the online world, such as the problems of Ferguson and Flint.[48] A trending topic in the digital media world is itself a social fact that must be grappled with; we might look to uncover whether the trending is partly explained by some kind of manipulation or algorithmic bias.

There is another point to be made about the Fergusons and Flints of the world. Once sustained attention was directed toward the issues, numerous new videos and pieces of evidence immediately flooded media platforms and hubs. Towns and cities facing similar problems were connected together. Once the network is primed and the symbols established—in this case, both hashtags and a shared knowledge and lens for generating meaning— information has the capacity to flow more quickly and freely. The common analogy is that a match can fall in the forest, but what happens next depends on whether the ground is wet from rain or bone dry. News media help prime networks for wildfire; they create a kind of ambient readiness for the spread of messages.

Most large networks are passive, with inert links between many individuals. Shared knowledge that is relevant to all changes the properties of the collective pathways. Suddenly, the network is activated, or utilized, and information begins cascading through. Further, by telling stories that evoke interest, empathy, and common understanding, journalists might generate the human currency and language (new symbols) to bridge a "cultural hole," a term coined by the sociologists Mark A. Pachucki and Ronald L. Breiger to explain the interplay between culture and networks. Networks are formed in a cultural context. Bridging across cultures, in all forms, is a vital function of news.[49]

In discussing the virtues of bridging, or "connection," Silicon Valley executives often miss the way in which news generates meaningful social contacts and group formation. "Whereas when you engage with public content," Facebook CEO Mark Zuckerberg has said, "you might get informed or be entertained, but it's not necessarily increasing social capital in the same way or building relationships between people."[50] This misses a subtle but crucial point. News, or quality public content, can generate conversation, which then drives more broadly shared social capital, not just individual-level capital. As Zuckerberg himself once stated, a "squirrel dying in your front yard may be more relevant to your interests right now than people dying in Africa."[51] Public content about, for example, "people dying in Africa" — that is, important news—is a necessary antecedent to generate meaningful social capital by stimulating broad discourse. We might normatively prefer that over the social capital generated by squirrel-related gossip among individuals.

An issue arises, and a policy is proposed. Journalists immediately look to find affected persons, members of communities whose experiences and perspectives might reflect the wider network of stakeholders. When news is impactful, it is because of the strength of formation of this public, which becomes powerful in its mutual awareness of shared problems and interests, their depth and texture. Knowledge is the key element in its awareness; it is what bolsters and builds ties that make up publics. Knowledge not only snaps the small-world nature of societies and body politics into place, but makes the small world visible to us all.

In certain ways, these ideas about a journalism of network building fit with the well-known ideas of *public journalism* or *civic journalism* promulgated by the journalism scholar Jay Rosen, who has long advocated for much richer engagement with communities and the issues articulated by citizens.[52] A movement in journalism that saw its peak in the 1990s, public or civic journalism has periodically seen renewed interest from scholars, who have called for a "second phase" in the social web era.[53] In any case, a network-building model of journalism also fits with other newer, innovative ideas about news, such as Rosenstiel's notion of seeing news as a form of collaborative intelligence.[54]

Dewey, writing nearly a century prior to the advent of social media, formulated ideas that powerfully explain the underlying dynamics of human communication relevant to this exploration. He observed that "symbols control sentiment and thought." Without the proper symbols, the "public will remain shadowy and formless, seeking spasmodically for itself, but seizing and holding its shadow rather than its substance."[55] Dewey noted that new "intellectual instrumentalities" are necessary to bring about social change. For networked publics to coalesce—for information to move through ties efficiently, and for ties to grow stronger—a symbolic context must be provided. The shadowy and formless public then transitions into a state of higher order. In an age of digital networks, the symbolic environment that allows such a new order to happen can be facilitated, in part, by the practice of knowledge-driven journalism.

3   Social Facts and Contested Knowledge

We should have no illusions that creating networks of recognition is easy. In many ways, the digital world is kicking up more sand in the gears, so to speak, for this kind of work. But that only means that what we might consider ethical "truth work"—what professional journalists are notionally committed to doing—is more necessary and vital.

Misinformation, false news, polarization, and general distrust make connecting communities more difficult. Algorithms and bots, or automated networks of fake accounts, can make these problems even more intractable, as they are less susceptible to human intervention. The latest scientific research continues to suggest that falsehoods and misinformation frequently beat out the truth—lies may spread faster and farther—in online networks.[1] We see daily headlines about the latest extreme digital trend heralding the end of civilization—and the increasingly poisonous, fragmented nature of Internet culture. Bringing together disparate persons around common issues can sound downright Pollyannaish.

Yet understanding exactly how knowledge is being contested—and how communities build different "civic epistemologies," in Jasanoff's phrase— is a first step. In the "post-truth" era, facts still remain, but it's easier to sidestep and circumvent them and find a like-minded interpretive community to build alternative meanings. And so facts matter—except when they don't.

To begin, let's take one of the more flagrant examples in the political realm in recent years. In the fall of 2017, Roy Moore was running for an open Senate seat in the state of Alabama. The firebrand former judge, a darling of the far-right, had cruised to a victory in the Alabama party primary. Despite grumblings from mainstream Republicans who preferred a more moderate candidate, Moore was expected to be elected by a comfortable margin. Then came a set of revelations and cultural-technological dynamics that created an utterly chaotic circumstance, highlighting the power of titanic new forces at work in the country's information ecosystem. On November 9, 2017, a *Washington Post* article alleged—in compelling, well-sourced detail—that US Senator Roy Moore had courted, groomed, groped, and assaulted numerous underage girls while serving as a thirtysomething district attorney.[2] Over the following days, more women came forward; their stories were credible and unambiguous, their motivations untainted by politics. Indeed, most were lifelong Republicans.

Sex scandals in politics are nothing new. The country spent a good portion of the 1990s embroiled in battles over politicians and sex. Yet those were set-piece skirmishes, the news cycles and the battle lines somewhat plodding and predictable. The candidate might have been forced to step down immediately, were such credible allegations to emerge. Or, more likely, such revelations would be swept under the rug, stuck in the perpetual purgatory of rumor.

But context is everything. Times had changed. The social networks of the country had been, in effect, wired and tuned just right to pay maximum attention. For weeks prior to the first revelations in the Moore scandal, the issues of sexual assault, abuse, and harassment of women were at the top of the mind for large portions of the American public. A cascade of revelations emanating first from Hollywood but extending across many professional domains had exposed powerful men such as the director Harvey Weinstein, the comedian Louis C. K., and the journalist Charlie Rose for their various abuses of power. Most of these revelations were years, even decades, old. Now they were exploding into view, one after another. A kind of electric sensitivity and desire for atonement had seized the public, fueled by social media hashtags such as #MeToo, which had galvanized the victims who had endured sexual harassment and assault.[3] They had bravely told their stories, by the thousands, on social media. Justice, too long deferred, had come due.

And sure enough, in the wake of the news about Roy Moore's pursuit of teen girls, an outraged country roared into action, filling Twitter and Facebook feeds with expressions of disgust and condemnation. The candidacy of Roy Moore had been hit by hashtags. It was deluged by informational crosscurrents and cascades from several directions, ones that had primed reporters and audiences alike to be more receptive to the allegations.

The candidate, however, possessed a powerful antidote: a grasp of the new social physics of information. Moore couldn't dismiss the women's stories, but neither did he need to step aside. Deep and conflicting forces in American culture had created a new possibility, one nearly inconceivable in any prior chapter in history: the facts would both matter immensely and not matter at all. It was as if the laws of news and facts had gone from a Newtonian universe to the land of quantum physics, where a particle can be both positive and negative, here and there, simultaneously.

The campaign's response to the stories revealed something new at work in the informational fabric of the nation. He immediately called the accusations a "prime example of fake news."[4] The framing of his denial was carefully calibrated; the words *fake news* were more than just a disparagement of the *Washington Post* reporters. They were a sharp signal and a call to arms, an appeal to a new sort of thinking, one that no longer had any respect or tolerance for, or deference to, mainstream institutions of news and information. Forces on the so-called alt-right had staked a lot on his candidacy. He was a cause for an insurgent group, despite the sitting Senate majority leader, a Republican, calling for him to step down. For these information warriors, the truth was irrelevant. But on December 12, 2017, a new topsy-turvy reality was made plain: Roy Moore was defeated in a bitter election by a Democrat, Doug Jones. Until a few weeks prior, the odds of a Democrat winning in a statewide election in Alabama, perhaps the nation's most conservative state, had been virtually zero.

The candidacy of Roy Moore—and of Donald Trump before him— unfolded against powerful forces of change, which had been building over decades. These same forces had scrambled public understanding of all kinds of issues, both policy and scientific alike. There was a central theme: Large portions of the public had grown to distrust the national media, for a variety of reasons. Many had begun to believe that most news is simply made up, fabricated. And it was of no help, surely, that actual fake news stories— elaborate hoaxes—circulated on social media throughout the 2016 election cycle, sowing confusion about what was real and what was fake.

What can be trusted in such an environment? Are the rules of truth now utterly, irrevocably up for grabs? Has an epoch of post-truth really descended? A new physics of news appears to prevail. As it turns out, however, the public has grappled with such questions before. In fact, the struggle has been going on, renewed with each successive wave of communications technology, for more than a century.

Fake News and the Iceberg

On April 14, 1912, the RMS *Titanic* collided with a massive iceberg in the North Atlantic, resulting in the deaths of more than 1,500 persons. This tragedy was a jarring event for the minds of early twentieth-century people across the Atlantic in many ways, some of which have been obscured by history. We think of it as a tale, almost allegoric in its power, about the hubris of modern engineering, the "unsinkable" ship mocked by the power and randomness of nature and the iceberg. Among other things, the sinking produced a long series of investigations that led to major changes in maritime regulation. And it certainly humbled modern society.

But lost in our collective memory and understanding of that event is how it also catalyzed the embryonic domain of telecommunications policy. More consequential than anything else that sprang from that historical event, the ship's sinking touched off a debate about the rules that would come to define media. Among the many regulatory consequences—an outcome largely forgotten—was the Radio Act of 1912, the first wholesale attempt by the federal government to regulate broadcast spectrum.

Investigators looking into the Titanic disaster determined that more ships could have helped survivors if their radios had been turned on at all times and if they used more robust communications practices. Although that maritime policy area needed to be remedied, the act also responded in part to the unfortunate fact, spotlighted by the press, that the broadcasts of amateur radio operators may have interfered in the ship's rescue, and citizen operators had purveyed, in the parlance of today, *fake news* of the *Titanic* being towed to shore.[5] With the Radio Act broadcast, licenses were universally required, and amateurs sidelined to shortwave signals. A headline of the day run in the *New York Herald* announced, "President Moves to Stop Mob Rule of Wireless."[6] The amateurs were considered "wireless meddlers," and public sentiment in the latter stages of the Progressive Era began favoring government control of the airwaves. Radio spectrum was to be considered something like a natural resource that needed to remain unpolluted.[7]

One hundred years of conflict ensued over who could speak in the public sphere, how much power they should wield, and, most of all, the role the government should play in determining the answers to such questions. The Radio Act of 1912 actually only gave the government the right to issue licenses. It would take fifteen more years, with the Radio Act of 1927, for the government to directly regulate radio frequency spectrum. These regulations ultimately created the groundwork for the Communications Act of 1934 and the birth of the FCC. Many factors helped contribute to government regulation of broadcasting, but the *Titanic*'s sinking certainly made the problem concrete and focused the early policy debate. Given that the event took place in international waters, the policy implications and subsequent debate over communications reverberated globally too.

That was a more innocent time. The answer was to stop misinformation by regulating and restructuring the transmission lines to the public. No such direct solutions are available in the twenty-first century. It would be trite overkill to portray Donald J. Trump as the ship of state ramming the proverbial iceberg, were it not the case that many Americans regard Trump's election as the greater tragedy. Taking the longer historical view, it is still political and nautical nighttime: the fog has yet to clear, the wreckage and the potential changes wrought are as yet undetermined. There is a broad sense that something in the communications ecosystem has gone terribly wrong, the signals of the public spectrum jammed and scrambled. Notions of a post-truth environment and a digitally mediated infosphere permeated by fake news have gripped public discourse.

It's become increasingly clear that this is no fleeting "moral panic" but a true, lasting crisis of fact and truth in the public sphere. We are also witnessing the embryonic stages of a new cycle in American communications-related politics and policy. There has been serious talk of further regulating the likes of Facebook and Google; the president himself has questioned the standing of news media outlets and even begun talking vaguely of their right to broadcast licenses.[8] It is the beginning of a long struggle from civic, corporate, and public policy standpoints to change the digital media ecosystem that now pervades our lives.

And like the *Titanic*, a historical event hasn't precipitated a public debate so much as crystallized it. Even before the world's most famous nautical disaster, Americans had been wrestling with problems of spectrum allocation and overlapping radio frequencies. Likewise, Trump's election made many people finally take notice of long-term media and communications trends. The election brought them to the surface.

In some ways, our apparent arrival in a post-truth environment—an era of impoverished understanding, or a "crippled epistemology," as political scientist Russell Hardin has labeled the mental world constructed by extremists—is unexpected.[9] Given certain trends, one would expect Americans to be better informed than ever. Citizens are, collectively speaking, better-educated, more inclusive, more connected, and more technically sophisticated than any previous generations. They have far greater access to knowledge and far greater individual capacity to express it. It is also a golden age of public scholarship, social science methods in journalism, and research-producing think tanks and organizations.

Despite this, or because of it, we find ourselves navigating the beginning of one of the greatest shifts in the realm of facts and knowledge since the Enlightenment cast its bright light over the epistemology of monks and priests. As with Gutenberg and the advent of the printing press, the digital revolution is changing fundamental structures of knowledge.[10] It is precipitating a crisis that transcends party or ideology. The past two former US presidents, opposed in party, philosophy, and policy preference, have each articulated the idea that there is palpable change in the information ecosystem. Former president George W. Bush has said, "Our politics seem more vulnerable to conspiracy theories and outright fabrication."[11] In a similar vein, former president Barack Obama has noted the dangers of fake news: "If we are not serious about facts and what's true and what's not—and particularly in an age of social media where so many people are getting their information in soundbites and snippets off their phones—if we can't discriminate between serious arguments and propaganda, then we have problems."[12]

It's increasingly evident that a great number of people—in America, in Europe, and across the world—cannot, in fact, discriminate between argument and propaganda and fact. What then must we do? The very first thing is to try to understand this new situation.

Social News: An Empirical Look

So what sort of news sees the highest levels of engagement on digital social networks? It's the kind of question that immediately elicits snickers, eye rolling, and mentions of cute cats and silly listicles—not to mention fake news and misinformation. Although many of the worst assumptions have a kernel of truth, the empirical reality is important to establish as more of the public sphere moves to online and socially mediated platforms such as Twitter and Facebook. Are we really "amusing ourselves to death" on Facebook?[13] People now spend an enormous amount of time with social media of various kinds. That kind of sustained attention may have real effects. What people are consuming on these platforms, in other words, has the potential to shape democratic debate and public opinion. In the wake of the 2016 election, many reports and articles have dissected the ways in which social media may have played an important, even decisive, role in turning the electoral outcome.[14]

There is a pervasive sense that somehow social media networks are fatally compromised as conduits for civically valuable information, and that it is a natural or inherent feature of social media platforms that superficial and/ or misleading content tends to prevail. This is, at root, an empirical matter, and we should test it. Can it be that social media promote behavior that overvalues the trivial? Is there something fundamental at work such that the democratization of news necessarily equates to the degradation of news?

Part of this public impression of superficiality might be accounted for by the sheer volume and velocity of news these days. News is being lumped in with tremendous volumes of other media content. Public affairs news is now appearing alongside gossip, rumor, infotainment, fake stories, and celebrity stories; news suffuses social media and is interpreted and filtered through countless visual memes and hashtags a day. The universe of "media" has grown, and a dizzying variety of media forms have converged in social media space together.

What precisely can we say about the news that is appearing in this space? The answer to this question speaks to any assertions of "inherent" qualities of social media—and indeed whether or not we might consider it even worth our time, as journalists and believers in quality public affairs news, to understand the deeper dynamics of networks.

It is certainly true that both the Obama 2012 and Trump 2016 campaigns successfully used data-driven techniques to "microtarget" potential supporters and to drive political engagement through social media. That issue of political campaign messaging, though important, is somewhat outside the scope of the analysis here; the question for our purposes is what kind of news media sees popularity and virality.[15] To examine whether social networks always race to the lowest common news denominator, I asked NewsWhip, an analytics firm that works with media companies and tracks patterns across social media, to share with me a half year of data from immediately after the election period, November 2016 to May 2017. The dataset, drawn from NewsWhip's API, was comprised of about seven hundred thousand stories; the selection was limited to the top five hundred news publishers on the web, as scored by Alexa.com, an Amazon-owned company that ranks websites based on traffic.[16]

The first thing to note is the substantial amount of what would be called *hard news*, which has been defined as "coverage of breaking events involving top leaders, major issues, or significant disruptions in the routines of daily life"; by contrast, *soft news* is "typically more sensational, more personality-centered, less time-bound, more practical, and more incident-based than other news."[17] Among the top one hundred news stories on social media as measured by Facebook engagements (like, shares, and comments) in the period immediately after the 2016 US election, about two-thirds were hard news stories. When examining a larger subset of the data, the top ten thousand stories from that period, the share of hard news overall declines but remains at around 55 percent. Further, using topic-modeling analysis to evaluate the most prominent themes among the top ten thousand stories, we see immigration, healthcare, Dakota Access Pipeline, Trump, Clinton, and other hard news topics rise to the top.[18]

If we move away from Facebook engagement as our chief measure of social media success and instead judge by Twitter engagements, nearly all of the top one hundred stories in the half year period immediately after the election were hard news in nature, and most were from outlets such as the *New York Times*, the *Washington Post*, CNN, the *Atlantic*, and Bloomberg. This difference between Facebook and Twitter suggests that there is nothing inherent about social media; the kind of news that thrives is likely a function of culture and the preferences of a particular user base.

As social media analysts have pointed out, several years ago sites such as BuzzFeed and Huffington Post were the big winners and leaders on platforms such as Facebook. However, as more traditional news publishers have adopted cutting-edge techniques and built their own social media teams, there has been a "flattening out" among the leaders (the top twenty-five publishers on Facebook) and there is less of a clear hierarchy.[19] For sure, viral soft news has seen a lot of success on social media over the past decade, and fake news had its moment during the 2016 election cycle. Both content categories will continue to command attention. But those do not necessarily represent the only major trends looking forward. Facebook has also continued to refine its algorithms, favoring quality news sources. The company can do more in this regard, and as it faces more criticism and further embraces its vital information role in democratic life, one can only hope its algorithms are tweaked to favor credible, ethical sources of news.

So why is it that social media gets such a bad rap? The answer may be, to some extent, in the outliers that see wide attention and shape views of social media as a whole. In the half year after the 2016 election, the top two stories on social media were the very definition of fake and soft news: first, with 1.9 million Facebook engagements, there was "Woman Arrested for Training Squirrels to Attack Her Ex-boyfriend," published by a dubious site called World News Daily Report; and second, with 1.25 million Facebook engagements, there was "Only People with Perfect Color Vision Can Read These Words," from BuzzFeed. Included in the top ten were also articles from other dubious publishing sites, with headlines such as "Female Legislators Unveil 'Male Ejaculation Bill' Forbidding the Disposal of Unused Semen" and "Angry Woman Cuts Off Man's Penis for Not Making Eye Contact during Sex." This kind of juvenile, often offensive humor and attention-grabbing fiction, which circulates among millions on Facebook, undoubtedly both entertains people and fosters a certain jaundiced attitude about the entire information ecosystem. Our collective views and judgments of social media may be formed—indeed, seared—as we encounter the inane and the ridiculous. Asked about their confidence in the news they encounter on social media, hardly any US adults (4 percent) say they have a lot of trust in that news, according to the Pew Research Center.[20]

Despite the circulation of a substantial amount of bizarre, trivial, and outright propagandistic content on online social networks, there is nothing inherent to social networks that makes them necessarily a haven for soft and fake news. Some research in this area has found that viral stories on social media tend to contain "overwhelmingly positive and awe-inspiring news," as well as news that has elements of unexpectedness or surprise.[21] Other research has even suggested that, for example, Twitter hashtags on politically controversial subjects are particularly persistent—their effects are sustained as people are exposed to them multiple times—and tend to cascade through networks more easily as compared to other, lighter kinds of fare.[22] Early adopters of many of the social networking technologies in question are generally younger, and the staffs of the companies in question skew toward the millennial generation. It is true that the algorithms used to detect and promote engaging content have been responsible for promoting false stories and stories with teasing headlines that leverage a "curiosity gap" (e.g., "9 Out Of 10 Americans Are Completely Wrong about This Mind-Blowing Fact.").[23] But it is also true that the user base—the public—determines the overall blend of hard and soft news by sending social signals that ripple across networks.

The empirical analysis presented here is only, of course, for one given period, and it is certainly a period inflected by heated political rhetoric and heightened attention to public affairs. What it shows, though, is that cultural shifts themselves can drive attention to hard news on social media. Hard news and public affairs content can thrive if the public is motivated and engaged by the issues. In fact, nearly 20 percent of the headlines among the top ten thousand news stories on social media from November 2016 to May 2017 mentioned the word "Trump" or variations of the same. It is not only the platforms themselves but also the citizenry using them that can drive attention.

Some of the latest empirical research is mixed on the relationship between the kind of news journalists typically supply, across all platforms, and what the public demands. It is a field that remains quite complex, and in some ways contradictory, in its collective findings. For example, in *The News Gap* (2013), Pablo J. Boczkowski and Eugenia Mitchelstein document the existence of a substantial divide in this respect: journalists consistently provide more hard news, such as politics and economics, than their audiences prefer.[24] This divergence narrows during certain periods—such as elections, which make politics more salient and drive more demand—but is consistent overall. By contrast, Joseph E. Uscinski in *The People's News* (2014) finds that US news production is substantially driven by audience demand, and news markets drive news content.[25]

News organizations surely still have some role to play in stimulating demand and interest and therefore bear some responsibility for what is going on in the media ecosystem. They can do much more to ensure their civically important content is engaging, to "optimize" stories and information for social platforms. Many news organizations, even the oldest and most venerable, are beginning to catch up in this game.[26] Some of this is as simple as testing headlines and images and making sure the initial "hook" for potential audiences is strong, relevant, and engaging.

More than two-thirds of Americans now get some of their news through social media channels, in which news is passed along and curated by friends and acquaintances.[27] What this signals is, in some ways, a new cycle in the history of communications and information, one that harks back to a kind of preinstitutional world where citizens would rely on sometimes underinformed, nonprofessional sources for information. Mass media still provides the vast majority of popular content, and large media institutions play an outsized role even within networks, as will be discussed in chapter 4. But there is little doubt that people's pictures of reality are being increasingly mediated and produced through more informal sources.

In a 1925 meditation on the role of newspapers in urban environments, sociologist Robert E. Park noted how mass media supplanted the more informal systems of passing along relevant or interesting information. "The first function which a newspaper supplies is that which formerly was performed by the village gossip," Park observed.[28] The personage of the village gossip, so to speak, has returned through social media platforms. Instead of the web embodying the image of the "universal library"—as it was conceived of in its early years—it has turned out that the web is much more like the offline world, where much of what we learn about and encounter is socially mediated.[29]

What might be the objective goal, then, of news media in a future of so much village gossip, with such information crosscurrents flowing? How might we even attempt to quantify success? How would a journalism that attempts to foster more overall knowledge on networks be evaluated?

One way of framing these problems is to imagine the public's information consumption and overall mindshare as one giant pie chart. What kind of information is spreading and radiating out through networks? We might imagine a three-way split among *cat viral* (infotainment), *catty viral* (polarizing content), and *civic viral* (content that engages citizens in public affairs.) The proportion of cat and catty versus civic viral might be estimated at about fifty–fifty. This is roughly in keeping with news production trends dating back to the 1980s; the overall share of hard news relative to soft news declined from 70 percent to 50 percent between the 1980s and early 2000s.[30] In the age of online networks, the goal of journalism now might be to increase the civic viral portion of the pie.

How Networks Are Changing Knowledge

The Internet has already brought major changes to the structure of all forms of knowledge, which have become more distributed and networked. This has both positive and negative dimensions in terms of fostering an informed public that values facts. In exploring these changes, we might consider what these new conditions might mean for journalists.

Let's begin with a general framework or hypothesis. In his 2011 book evocatively titled *Too Big to Know: Rethinking Knowledge Now that Facts Aren't the Facts, Experts Are Everywhere, and the Smartest Person in the Room Is the Room*, David Weinberger observes that there has been a "thorough change in the shape of our knowledge-based institutions." Broadly speaking, Weinberger writes, the distinguishing qualities of this new epistemological paradigm are that knowledge is increasingly "wide" (as in crowd-sourced), "boundary-free," "populist" (or nonhierarchical), "other-credential," and "unsettled." He sums this up as follows: "It's the connecting of knowledge—the networking—that is changing our oldest, most basic strategy of knowing."[31]

According to Weinberger, the idea of a foundational set of data and information upon which highly reduced and synthesized knowledge and wisdom rest—the famous DIKW pyramid model outlined in 1988 by organizational theorist Russell Ackoff—increasingly is no longer valid. (In the DIKW pyramid, the layers from bottom (largest) to top (smallest) are data, information, knowledge, and, at the pinnacle, wisdom.) The Internet provides a networked model, not a foundational one, for knowledge. And the net's near-infinite capacity for networks, nesting, hyperlinking, and variable algorithmic recommendations of related information make the age-old demand for reducing what is known to its essence no longer a necessity for humans.[32]

One can quibble over whether this is overstating the phenomenon, but it furnishes a useful framework with which to analyze current and future trends. For sure, the intellectual world is not flat across every domain. Work by a number of scholars, such as Andrew Chadwick and James Webster, suggests a "hybrid" or "overlapping" set of cultural and media ecosystems, in which powerful gatekeeping institutions still play a strong role even as news becomes more collaborative and information more democratized.[33] Further, networks can have their own hierarchies, as David Grewal has pointed out. (This is a point I will explore much further in the context of network science, in chapter 4.)[34] These are, in any case, relatively early days for the Internet. This landscape is likely to evolve in the direction Weinberger suggests, given the dramatically lowered costs of publishing on the web.

It is useful here to return to Jasanoff's idea of civic epistemologies. The way that publics come to know complex phenomena—think climate change, health care, cybersecurity—cannot be explained by "the sum total of a population's understanding of a few isolated scientific facts." Rather, there are always deeper *knowledge-ways*, forms of "collective knowing" and shared approaches to sense-making that guide public knowledge. Jasanoff's idea is that these forms of knowing are very deeply embedded in cultures, and experts and laypersons alike are bound up in them.[35] All of this is true, irrespective of ICTs or the web. Yet what digital networks may allow for is a proliferation of such civic epistemologies, as large groups are able to organize knowledge-ways that stand as alternatives to culturally dominant belief structures.

The main idea to consider is that knowledge may look increasingly flat and unfinished in many domains—and it is being updated constantly. This is a function of what the technologist Nicco Mele has called a general condition of "radical connectivity."[36] ICTs allow for the widening of knowledge outward, on every conceivable subject. What was once a societal demand and tendency toward hierarchy and reduction has become more about process, about endless iteration. Mass media no longer control what is deemed valid or valuable. What does this mean in practice? Do particular facts matter anymore, or is it just one big conversation, everyone with her own perspective? Facts, I hope, do still matter—but the ever-expanding digital world allows for infinitely more context and new forms of knowing. Let's investigate what this looks like, the good and the bad, and look squarely at the emerging reality that we face.

Popular Delusions and Social Facts

When historians look back on our post-truth era, it is likely that one particular fact may be seen as the very emblem, the synecdoche, of all the shifts in the ground floor of news and information: the fact that President Barack Obama was born in Hawaii, but sizable portions of the public continued to believe he was born outside the United States. No amount of fact-checking by the press, nor the release of his actual birth certificate by Obama himself, could quell the so-called birther rebellion against what was, indisputably, an established fact. Time and again, just as the lie was about to die, the rumor that Obama was born on foreign soil, thus making him ineligible to be president, was kept alive by websites and the digital ecosystem as a whole. The real birth certificate was endlessly dissected online and declared a hoax; hoax Kenyan birth certificates were circulated; elaborate conspiracies threaded their way through millions of social networks.[37] On one level, this was a case of partisan forces pushing a lie—not new in political history—but digital networks now give this kind of social fact unprecedented scale, speed, and persistence.

It seemed a silly distraction at some moments. But the birther issue, it should be remembered, also propelled Donald Trump back into the national spotlight, giving him an initial foothold in the roiling Republican debate about how to take back the White House from the Democrats. In a weird way, the manufactured birth certificate controversy was both utterly trivial and totally consequential, insofar as it motivated and consolidated deep societal forces that surely were inspired in part by xenophobia and that surfaced in the 2016 election.

Polls taken as late as 2016 showed that perhaps 40 percent of all Republicans still doubted President Obama was born in America. Further, polls that differentiated between lower-knowledge persons and higher-knowledge Republicans (established by pollsters through asking basic civic questions such as how long a senator's term is) showed that more politically knowledgeable persons were no less likely to believe the conspiracy. "A greater factual understanding of the political system does not diminish Republicans' doubts about Obama's birthplace," stated an NBC News report on its polling findings. "The fact that more Republicans currently think that the president was not born in the U.S. and that this belief does not depend on how knowledgeable they are about politics is surprising. The country may be divided about both facts and opinions."[38] It is at this point we have to ask what was really going on. Was the denial of President Obama's origins just racism, or some sort of coded protest vote? The reality is difficult to assess, but we might look to history for perspective.

History is replete with instances of mass hysteria over falsehoods, rumors, and illusions. The canonical example is the Salem witch trials, but the phenomenon repeats itself in nearly every age, across global societies. A book published in 1841, *Memoirs of Extraordinary Popular Delusions and the Madness of Crowds*, by Charles Mackay, details the extraordinary range of sheer folly that humanity has engaged in, including belief in the supernatural, magic, pseudoscience, and overheated rumors of every kind about some matter of economics and politics. As Mackay notes, "We find that whole communities suddenly fix their minds upon one object, and go mad in its pursuit; that millions of people become simultaneously impressed with one delusion, and run after it, till their attention is caught by some new folly more captivating than the first ... Men, it has been well said, think in herds; it will be seen that they go mad in herds, while they only recover their senses slowly, and one by one."[39] The herd instinct—the madness of crowds, in other words—is well established as a phenomenon. The Obama birth certificate issue is just one recent example.

Social scientists have been examining the phenomena of what is sometimes called *motivated reasoning* or *biased assimilation*.[40] Humans filter facts through worldviews, belief systems, and the social communities that support these systems. It is by now a well-documented pattern, going back many decades in the field of social psychology, that social cues and social pressures even from strangers exert a powerful effect on individuals' judgments.[41] Moreover, accepting new facts and beliefs can threaten one's group affiliation. Faced with (1) accepting the fact that climate change is being driven by humans and enduring ostracism from friends who believe the contrary or (2) denying the fact of climate change and maintaining group identity, humans will often prefer the latter: to keep their community and say to hell with the facts. Further, once individuals identify someone as sharing the same political worldview, they will tend to trust that person on a wide variety of topics beyond politics. (This is called an *epistemic spillover*.)[42]

People will often choose to reinforce their connections to others with whom they share a worldview, rather than risk breaking those connections. What is particularly intractable, and somewhat frustrating, about this problem is that education and knowledge levels seem to have little effect. In fact, better educated individuals sometimes exhibit even greater resistance to scientific facts; on an issue such as climate change, there remains a great divide among educated individuals.[43] Indeed, the evidence is that most people who believe in conspiracy theories are far from crazy; they are often quite rational, with a broad command of relevant facts, but they have surrounded themselves in a network of like-minded persons who reinforce their beliefs.[44]

Conspiracy theories can, under certain conditions, lead to extremism and fanaticism, which Hardin notes is inherently a "sociological and not merely a psychological matter."[45] Fanaticism grows when "constrained belief systems" isolate a given group; this can happen "any way that cripples knowledge of the groups' members," although it typically involves a small leadership cohort that enforces discipline and cohesion.[46]

In any case, if popular delusions and stubbornness about conspiracies are eternal phenomena, what has changed for us, now, in this current time and place? For one, ICTs allow for cascading behavior—mass peer-to-peer sharing—of conspiracies and half-truths much more quickly. These have been called *availability cascades*, whereby "expressed perceptions trigger chains of individual responses that make these perceptions appear increasingly plausible through their rising availability in public discourse."[47] Rather than rumors quietly passing from village to village—as they would have in Mackay's nineteenth century—the spread of such cascading phenomena is now utterly visible in real time on social media, creating its own kind of reality and producing feedback loops that provide fuel for conspiracies.

Hashtags related to the birther conspiracy became hard to ignore once they reached the hundreds of thousands or millions of persons who seemed to be paying attention to or reinforcing the idea or belief. Once many persons claim to "know" something by retweeting or liking it, that "vote" on digital platforms itself becomes a fact. Innumerable news stories about the Obama birth certificate conspiracy were based on just that: they were marveling, quizzically, over the very existence of such widespread belief. Social media provide a kind of constant stream of quasi-public opinion data, even if a given self-selected group believing a conspiracy is often wildly at odds with what majorities believe.

A further iteration of this centuries-old problem is the capacity of the web to allow group formation among geographically disparate persons into communities that exhibit *epistemic closure*—hard-core communities of corrupted epistemology that tend toward extremism. Such communities, as Kate Starbird notes, are not entirely cut off from news media and fact-checking efforts, but they tend to use media denials as further fuel. To media corrections, they say: "See, the media must be trying to suppress the truth." Starbird's research suggests whole ecosystems of websites frequently sprout around, for example, breaking news events. These conspiratorial sites replicate similar content, giving the impression of diverse sources documenting the same (bogus) information, such as that the government staged a given school shooting. These "alternative media domains may be acting as a breeding ground for the transmission of conspiratorial ideas."[48]

It remains unclear how much professional journalism should respond directly to deliberate misinformation. Take, for example, a rather trivial example, but one that exemplifies the kind of item that spreads almost every hour across social media. On February 22, 2018, a rumor began spreading that First Lady Melania Trump had hired an exorcist to perform a ceremony at the White House. The message got virtually no attention until PolitiFact, which has won a Pulitzer Prize, decided to debunk the rumor. "NEW: No, Melania Trump didn't hire an exorcist to cleanse the White House of previous administration's demons," PolitiFact's Twitter account announced. It was at that point that discussion began to kick into gear on social media. The silly rumor didn't ultimately go very far, relatively speaking. But according to an analysis by TwitterTrails, a project run by Wellesley College computer scientist P. Takis Metaxas, it was PolitiFact's intervention that gave the rumor whatever viral lift it had.[49]

Media repetition can give fuel to conspiracies and even backfire, sometimes confusing the public, whose memory of the corrected record and actual facts can fade quickly.[50] The journalism and communication scholar Lucas Graves has documented the tensions implicit in the fact-checking movement itself. Professional fact-checkers sometimes try to assert strong authority even when issues may have interpretive nuance or reasonable ambiguity, and decisions to select and fact-check certain claims, and not others, may be fraught with journalistic value judgment.[51] This may be overcome to some degree by being transparent about uncertainty and nuance. Using ratings scales to indicate degrees of truth may make corrections more effective in some cases.[52] In any case, fact-checking remains a noble and necessary undertaking, but it is insufficient to solve larger problems that stem from group polarization and availability cascades.

Finally, relevant to the concept of social facts is a psychological mechanism that social scientists have labeled *asymmetric updating*. This theory predicts that good news and bad news on issues, even objectively and dispassionately delivered news, will cause entirely different effects based on news consumers' prior beliefs or antecedent convictions.[53] Asymmetric updating means that an accurate, objective news article—for example, one exploring a study that shows climate change either slowing or accelerating—may not do what is intended. The very antidote to polarization and bias that is traditionally promoted—better, more accurate, more balanced news that could anchor reasoned debate and the public sphere—may produce the very opposite of the desired result.

There is indeed a long series of research findings going back decades that would lend support to this kind of phenomenon: that balanced news might increasingly produce "unbalanced" views.[54] The social science literature on belief formation has a number of related theories, including biased assimilation and motivated reasoning.[55] The general phenomenon of people interpreting and selecting facts differently based on the same body of evidence and updating their beliefs accordingly is well established.

Research continues to find that exposure to Internet news can lead to increases in polarization, although these effects may be somewhat modest in the short term (what most experiments can measure).[56] This is true despite the fact that some recent empirical findings suggest that it is the group least connected to social media and web-based media that has shown the greatest degree of polarization.[57] A reasonable hypothesis is that as more of the news population is polarized through partisan media and builds up beliefs, or "priors," on a range of issues, this affects how consumers regard professionally produced news.

Representing Networked Knowledge

To the traditional, canonical notion of "knowledge"—systematic information that can be validated in a scientific sense, independent of individual or group judgments—we are fast adding the parallel domain of knowledge of social facts. Again, these are the visible body of mass reactions to, perspectives on, and behaviors regarding any given fact, event, or phenomenon. Through social media platforms, our visibility into the beliefs of the crowd itself is producing another category of relevant empirical knowledge. Yes, it is an objective fact that Obama was born in the United States. Still, it is an objective fact that large numbers of people believe (or believed) he was not. This is the dilemma and paradox now faced by all institutions that claim authority over knowledge, news outlets especially. To ignore the mass delusions and deliberate, partisan falsehoods of the online world is in some ways to ignore the facts—the social facts.

To formulate a model that might help us see these changes more clearly, we might come back to Weinberger's idea that knowledge is increasingly networked (see figure 3.1). We might see these changes as representing a shift from knowledge conceived as a small-world network—highly connected nodes, built around conventional credentials of authority—to a very large, "scale-free" network, with heterogeneous nodes and several highly connected hubs that stabilize the structure at large. In this sense, a *node* is some sort of authoritative source, bit of knowledge, or perspective that is valued by others in the network, as represented by mentions, retweets, hyperlinks, or other connections.[58]

In this shift from a hierarchical to a flat (populist, networked) representation of knowledge, what has really happened is that the knowledge (universe of nodes) itself has not really changed. What has changed is the structure and organization of that knowledge (again, of the nodes). Specifically, this has happened via destabilization caused by disruptors, such as social media or technology in general. Whereas the structure was initially a small-world network, what we now have is a very large, very sparse network that behaves seemingly randomly in its spontaneous reorganization

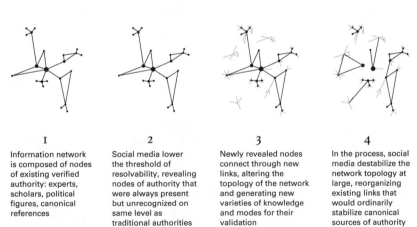

I	2	3	4
Information network is composed of nodes of existing verified authority: experts, scholars, political figures, canonical references	Social media lower the threshold of resolvability, revealing nodes of authority that were always present but unrecognized on same level as traditional authorities	Newly revealed nodes connect through new links, altering the topology of the network and generating new varieties of knowledge and modes for their validation	In the process, social media destabilize the network topology at large, reorganizing existing links that would ordinarily stabilize canonical sources of authority

Figure 3.1
How social media and ICTs change knowledge structures.
Source: Steven Braun and John Wihbey.

whenever something (e.g., a viral tweet) disrupts or destabilizes it. Of course, that very large, very sparse network has always been there, and what has changed is not the network itself. The threshold of detectability of its nodes (i.e., the infinite nodes of information and knowledge that would ordinarily be discredited by figures of conventional authority, in a hierarchical representation of knowledge) has been lowered. Put another way, the resolution at which those disruptors operate has radically increased.

We might call this lowering the *threshold of resolvability*. In this case, *resolvability* refers to the ability of the network to be broken down into constituent parts. It also refers to detectability: more nodes are visible because they are now detectable in a way that they were not before. Specifically, social media make more nodes visible simply by way of giving them a platform to be heard and disseminated outwards. (The term *resolvability* is borrowed from the natural sciences.) ICTs and social media lower the threshold of visibility and detectability, or *resolution*. It's like working with a microscope: without the microscope, only the largest growths on a cell plate are visible, but with the microscope, so much more comes into view, and the view becomes flattened in the process.

Social facts, with their accompanying attention signals, make visibility into disparate network nodes, clusters, and knowledge communities much easier. Algorithms, picking up on the data of social facts, can help drive this visibility into the corners of a scale-free world. Within social online communities, certain civic epistemologies, or pathways to knowing and validation, can help lower the threshold of resolvability. If one's social group is predisposed to doubt the need for vaccines or to address climate change, for example, it becomes ever easier to find more information that confirms one's biases.

This model of resolvability represents a rather general claim about the evolving state of knowledge in the world. Its validity and explanatory power will vary across domains, with a wide spectrum of effects. Obviously, issues of policy and values are susceptible to vast proliferation of social fact creation, of communities that together interpret facts in novel ways. Yet science, too, can be subject to the same forces. The issues of vaccines and climate change may continue to see substantially splintered networks with their own civic epistemologies. Institutions that produce knowledge will continue to interact with individuals and groups in dynamic and unpredictable ways, sometimes anchoring communities more in empirical fact while failing to do so at others.

Strains of Information Disease

So what do all of these trends, both cultural and technological, portend for the future of public knowledge and its structure?

It is not just the specter of misinformation that haunts our moment and creates anxiety: it is also, paradoxically, the very presence of abundant and free information. In the hurricane gale of noise online, there is a sense that the truth doesn't stand a chance. Technology and legal scholar Tim Wu has suggested that the First Amendment may be obsolete, but not for the reasons we might think. He notes that "if it was once hard to speak, it is now hard to be heard. Stated differently, it is no longer speech or information that is scarce, but the attention of listeners."[59]

Digital technologies are accelerating the war for attention; paired with algorithmic targeting techniques, these communications technologies can contribute to echo chambers and social fragmentation. The possibility of constructing "shared realities" for citizens to have deliberative discourse over issues of substance, with a modicum of shared facts and values, therefore becomes less likely. There is an interrelated and inexorable trend that compounds the problem of fragmentation: personalization. In one respect, the personalization of news and information media products has the potential to better serve individuals' specific needs and tastes. But personalization—which is fast becoming the watchword of our age, from medicine to education to entertainment—also potentially erodes the time-honored concept of a public sphere.

Legal scholar Cass Sunstein, who has been sounding the alarm about this issue for nearly two decades now,[60] asserts that a "fully personalized speech market, consisting of countless niches, would make self-government less workable."[61] A world of radical information choice can theoretically jeopardize the material necessary for democratic consensus. "In important ways, it would reduce, not increase freedom for individuals involved," Sunstein writes. "It would create a high degree of social fragmentation. It would spread falsehoods, some of them dangerous. It would make mutual understanding far more difficult among individuals and groups."[62]

Yet the problem is not just fragmentation itself, but the lack of account-ability that data-driven microtargeting of media and messages affords. The various specters of fake news, false news, misinformation, and propa-ganda—of rumor, gossip, half-truth—have always been present in human societies. The difference now is that manipulation may be hidden almost entirely from public view and scrutiny, as illustrated by the Russian-backed propaganda efforts through the Facebook platform during the 2016 US presidential election. Such capabilities radically reduce the ability of report-ers to play their traditional watchdog role. "In a world of micro-targeted messaging, it is difficult for reporters and scholars to know who is saying what to whom, where and with what effect," political scientist Kathleen Hall Jamieson notes. "In the absence of such information, journalists' abil-ity to hold sources accountable is even more circumscribed than when pseudonymous groups broadcast their messages in places open to public view."[63] In the marketplace of ideas, more sunlight and more speech are the traditional antidotes to corrupt interests and "bad speech" in a democracy. Yet countervailing speech and transparency do not have the same power in the face of such new media technologies.

The general information disease that has been called *fake news* in some ways distracts from a deeper structural pattern: the decline in public trust in institutions across many domains. The disease of distrust is particularly acute when the public is asked about news media. Ill-intentioned actors make this problem worse. Jeff Jarvis comments, "'Fake news' is merely a symptom of greater social ills. Our real problems: trust and manipula-tion."[64] There will be no easy solution to these intertwined issues, partic-ularly in a time of media choice and fragmentation. When asked about whether they trust the news outlets they follow, members of the public still generally say yes, particularly if the news outlet (e.g., MSNBC or Fox News) conforms to their political ideology.[65] It's the stuff they don't follow—and assume is rubbish—that they distrust implicitly; and these attitudes pro-duce declining evaluations for news media in general.

Virtues of Knowledge Decentralization

So far our story about knowledge has not been a happy one. It seems that digitally networked platforms only stand to accelerate the worst in human nature, to make visible an ugliness and an irrationality that has always been with us but was, until the recent past, somewhat moderated by authorities and anchoring institutions such as governments, courts, universities, libraries, and expert communities that could arbitrate on questions of science and knowledge.

Yet decentralization and its defects are highly context-dependent. It is worth remembering that perhaps the most influential early argument for a decentralized, nonhierarchical system of knowledge came not in radical form, but sprang from the mind of Nobel laureate Friedrich Hayek, an economist and philosopher whose book *The Road to Serfdom* (1944) stands as one of the classic arguments against both fascism and socialism—and state central planning of any kind. In a paper titled "The Use of Knowledge in Society," published just after his classic book appeared, Hayek noted that no centralized system of economics can beat, in a sense, the wisdom of the crowd.[66] The real knowledge about notions of value and preferences for prices are radically dispersed across a society; the free-market price system of capitalism solves this problem of efficiently aggregating dispersed knowledge better than any centralized system that fixes prices.

Hayek, importantly, did not think that all forms of knowledge are better off sourced from the crowd. "It may be admitted that," he writes, "so far as scientific knowledge is concerned, a body of suitably chosen experts may be in the best position to command all the best knowledge available—though this is of course merely shifting the difficulty to the problem of selecting the experts." But there are classes of problems and domains that stand outside of the realm of formal testing and expert methods. Hayek notes that a "little reflection will show that there is beyond question a body of very important but unorganized knowledge which cannot possibly be called scientific in the sense of knowledge of general rules: the knowledge of the particular circumstances of time and place." Knowledge of social, geographic, and lived reality is something that must be harvested from distributed sources to discern a pattern. The crowd is not always wise individually, or smart even in its majority expression, but its inputs are essential on some questions. Jimmy Wales, who founded Wikipedia, has said that Hayek's ideas helped influence his vision for a collaborative, online encyclopedia.[67]

Information systems that are decentralized do jobs that are important, jobs that pure science and expert systems cannot do well. Further, there is an emerging idea in social science that should give us pause before we disparage the social world too much as a source of knowledge. As Steven

Sloman and Philip Fernbach explain in their 2017 book *The Knowledge Illusion: Why We Never Think Alone,* the majority of what we think we know is not stored in our heads. In fact, humans retain shockingly little detail about anything. The failure to recognize this deficit leads to all sorts of bad outcomes, such as overconfidence in ourselves and a failure to realize how much others affect our reasoning. But knowledge acquisition and production are an inherently social act. We all live in a "community of knowledge," relying on fellow human beings for almost everything we need to do.[68] Sloman and Fernbach note, "Individual intelligence is overrated. ... We learn best when we're thinking with others." Their point is that we should neither credulously believe what our communities tell us nor take for granted what credentialed experts say. What we should do is more judiciously choose the communities we affiliate with.

Although the dark side of the Internet is its acceleration of misinformation and antisocial behavior, the other side of networked technology is that it creates the possibility for increasing collaboration to answer shared questions. "The power of crowdsourcing and the promise of collaborative platforms suggest that the place to look for real superintelligence is not in a futuristic machine that can outsmart human beings," Sloman and Fernbach observe. "The superintelligence that is changing the world is in the community of knowledge."[69] Technology, then, provides for both greater efficiency in the aggregation of lived experience, as Hayek notes, and greater capacity to build large groups that can solve problems.

The dream of using technologies in this more prosocial and proknowledge way—as what are sometimes now called *knowledge networks*—goes back a long time. Some fifty years ago, the visionary technologist and psychologist J. C. R. Licklider, a key figure in the Internet's early development, laid out a remarkably prescient (and optimistic) blueprint for building communities of knowledge—for the application and advancement of knowledge—in a book called *Libraries of the Future.*[70] In it, he presented a vision for "procognitive systems" whereby the thoughts of people, as well as formal knowledge from printed texts, could be organized and brought to bear to help work on questions. This "fund of knowledge," as he called it, could be called upon conceptually, across disciplines, based on semantics and meaning, not just tagged information retrieval as in traditional library science. Humans and machines could work together toward knowledge development. The date that Licklider fixed on for the development of such a technological system was the year 2000.

Licklider's vision was conceived in the "heady optimism of artificial intelligence at the Massachusetts Institute of Technology (MIT) and surroundings in the early 1960s."[71] Now that the sort of natural language processing and semantic, topic-oriented computational capacity is here with us in the twenty-first century, we are closer, at least technologically, to achieving this more positive vision of an online social community. His book's final, speculative chapter, "An Approach to Computer Processing of Natural Language," stood as a largely unrealized dream for half a century.[72] Now it is here. Whether or not such technological tools are realized for such grand purposes, however, remains to be seen.

A Nation of Googlers

The availability of knowledge through the click of a button has also begun changing how individuals evaluate credibility and authority. This is an area that requires more study, but certain patterns are emerging. Alison Head of Project Information Literacy has taken surveys with thousands of college students in recent years and concluded that in terms of the bar for credible sources, there is a dramatic drop in expectations from truth to consensus among college-educated youth.[73] This is not wholly negative; students are using social communities online to help point them in the right direction. Students report being guided as much by blog comments and social media conversation threads as any traditional markers of authority.[74] They are looking for the "wiki voice," the back-and-forth, colloquial threads that seem to cascade along forever on interesting topics. It's the human Greek chorus, which, at its best, represents others helping others—what Head has called the concept of "shared utility."[75]

In terms of knowledge access, our new world relies tremendously on search engine algorithms to deliver the most useful and credible sites (not just the most popular or entertaining) on the first results page. Yet the algorithms themselves are not always well designed to do this. Further, there are substantial changes in terms of how deep many people will go in trying to track down the best possible version of knowledge. In our post-truth world, the evaluation of knowledge may have become a more perfunctory process facilitated by the ease of the one-search interface. Many of us, not just students, have become a nation of Google searchers looking for quick matches of facts and figures, rather than interrogating the credibility of the information we find online and reflecting on how it informs our thoughts, beliefs, and opinions. Knowledge has become networked, and this has had consequences for how we interact with knowledge.

Yet it is worth thinking about networks of peer-produced knowledge—blogs, social media communities, and so on—as more than just a rabble of amateur noise. They may be seen as a hybrid adaptation to technology in an increasingly complex world. Here's one way of thinking of this revolution: we are seeing the blossoming of new kinds of organic or biological systems, or "knowledge ecologies," as Bill Cope and Mary Kalantzis call them. "More knowledge is being produced in the networked interstices of the social web, where knowing amateurs mix with academic professionals, in many places without distinction of rank," they note.[76]

Yochai Benkler points out that the "networked public sphere" has developed useful features of accreditation and filtering "without re-creating the power of mass media." We should be reminded that, in a different kind of way, knowledge and claims about the world do have a system of checks and balances, at least in some web-based communities. Benkler outlines ideal systems as a multilayered process of commenting and claim-vetting: "'Local' clusters—communities of interest—can provide initial vetting and 'peer-review-like' qualities to individual contributions made within an interest cluster. Observations that are seen as significant within a community of interest make their way to the relatively visible sites in that cluster, from where they become visible to people in larger ('regional') clusters. This continues until an observation makes its way to the 'superstar' sites that hundreds of thousands of people might read and use."[77] This is not the same as academic peer review, obviously, but it may be better in many cases than the review process through which a news organization evaluates information, in which only a general assignment reporter and editor may be involved.

Media Framing and Public Knowledge

As we pay more attention to instances of misinformation circulating online—and of sometimes strange and wild social facts bubbling to the surface—we begin to see them everywhere, every day. How troubled should we be? Yes, a doctored image of a natural disaster, a fake Twitter post, or a bogus name attached to a mass shooting can all serve to unsettle the public, making it incrementally harder to discern fact from fiction. Social media companies and search engines can make this worse by allowing algorithmic systems to feature such content prominently. User-generated content is hard to verify; journalists sometimes make it worse by rebroadcasting such content.

Yet, as mentioned, we should acknowledge that hoaxes, conspiracies, and fake news have always been with us and always will be. Just look at the magazine rack in the checkout line at the grocery store. As lamentable as fake news and deliberate falsehoods are, to obsess over them misses massively more consequential phenomena.

When historians look back, what will be the most consequential media-public information disasters of our time? The mass incarceration crisis in the United States and the country's failure to address climate change are two prime candidates, for sure. Both are the product of erroneous cultural narratives, not particular pieces of doctored or falsified content. Neither has much to do with fake news, conspiracies, or hoaxes. Both also have to do with misguided forms of what social scientists call "framing," or the kinds of words, images, and context that journalists choose that then serve to produce certain kinds of meaning and narratives in the minds of the public. *Media framing* has been defined as a "central organizing idea or story line that provides meaning to an unfolding strip of events."[78] This should remain the area of central concern in our media environment, as the effects can be long-lasting.

There is a growing awareness that the phenomenon of mass incarceration constitutes a moral crisis in the United States. More than two million persons are now imprisoned, many of them racial minorities who became involved in the prison system at a young age. Liberals and conservatives alike have increasingly acknowledged this as a catastrophic error in public policy.[79] There is a fairly good case that a media-driven feedback loop, beginning in the 1980s, helped produce this situation. News media sensationalized violent crime through the early 1990s, fueling public fear and providing a distorted picture of risk for society. New laws were passed mandating draconian sentences. With the amount of crime-related coverage roughly tripling between 1992 and 1994 alone, there is little wonder that politicians were clamoring to pass new laws. *Time* magazine's cover story in 1994 said it all: "Lock 'Em Up and Throw Away the Key: Outrage over Crime Has America Talking Tough."[80]

It is noteworthy that "America" was reacting to "outrage," to a perception, not to data or reality itself. Remarkably, new laws and tough attitudes came about even as the rate of violent crime was already dropping overall. Society's approach to crime backfired because, in part, news media coverage was disconnected from knowledge. The policy consequences were devastating. Further, new research is beginning to show that what journalists should have been covering more of was right in front of their noses—innovative community policing approaches, often rooted in new nonprofits and citizen groups. These were the real solutions, lowering crime and keeping young persons out of prison. Citizens taking action had a big effect where there were problems.[81] These approaches were seldom covered; news media instead opted for lurid headlines and salacious images. Had more networks of recognition been facilitated, perhaps the outcomes would have been different. Now we live with the wreckage, and communities of color continue to see profound effects.[82]

Despite the many public lamentations now of the sequence of events that led to the mass incarceration crisis, one central underlying driver—exaggerated and distorted news attention—remains constant. A study conducted from 1998 to 2002 of 2,400 local television newscasts, randomly sampled across markets, found that 24 percent of stories were crime-related, and 61 percent of newscasts led with crime, disaster, or accident stories. A 2005 study found that this "it bleeds, it leads" or "hook and hold" strategy in local television was accelerating, with 77 percent of all first newscast stories relating to crime, disaster, or accidents.[83]

This pattern seems to be continuing. As part of our ongoing Reinventing Local TV project at Northeastern University, we recently coded 1,061 stories from the leading stations in each of the top fifteen markets across the country (according to available ratings data in early 2017).[84] In this week's worth of newscasts, we found that 28 percent of stories were crime-related. Many of these stories featured racial minorities who are frequently overrepresented, even accounting for their higher representation in urban areas (where the large TV markets are), in news about crime, despite slight improvements by news channels in recent years.[85]

The second great news-driven crisis we might identify is the American public's response to the problem of climate change, which has been underwhelming at best. The United States is now the only country in the United Nations to oppose the Paris Agreement to limit greenhouse gas emissions. How did such a bizarre situation arise? Again, no conspiracy or hoax or bit of fake photography alone produced this situation. Rather, news media over many years generated a narrative suggesting that the reality of man-made climate change was plagued by uncertainty. Deniers, skeptics, and energy-industry-backed "merchants of doubt," as Naomi Oreskes and Erik Conway memorably labeled them, were given the megaphone by main-stream outlets.[86] Only when it became too late did professional news out-lets begin to move away from the framing model of *false balance*, or giving equal weight to both sides despite the scientific consensus.

As much as bits and pieces of misinformation circulating on social media may have impacts here or there, nothing will continue to matter as much as large media outlets "getting it right," so to speak, and establishing pub-lic framings and narratives that are informed by knowledge. A doctored picture of a shark swimming down a flooded street has virtually no policy implications; nor does almost any rumor or piece of information online, until it is picked up and legitimized by a large media organization. What matters is that the narrative-shaping institutions do a better job, following the data while also being aware that statistics often leave out emerging pat-terns and new information. Fake news is not a red herring, but we need to both fight fake news *and* improve "real" news.

The psychologist Jerome Bruner argues that humans organize their expe-riences and construct their model of reality primarily in the form of nar-rative. Narratives are a "version of reality whose acceptability is governed by convention and 'narrative necessity' rather than by empirical verifica-tion and logical requiredness."[87] Narratives are the way the public comes to knowledge. In a digital society like ours, the conventions and symbols that constitute stories, which allow the public to fit in selected experiences, are substantially facilitated by news media.

The engineers of Facebook and Google and teams of fact-checkers play-ing whack-a-mole with fake news items can only do so much. We should pay careful attention to factors that may have a more consequential role in shaping public understanding. Even in a chaotic, social-media-driven world, the direction and quality of real news matters. The next societal blunders on par with the mass incarceration and the climate change inac-tion crises are most likely to be produced, or substantially abetted, by jour-nalists, not fake news purveyors. Journalists can substantially help shape the narratives and frames that develop within networks, even if they can-not control the precise outcome.

4 Understanding Media through Network Science

So knowledge is now distributed, sewn in a million places in digital space. The crowd and the "library"—the repository of knowledge as produced by traditional institutions—coexist as competitive partners across many topics. Facts produced by authoritative sources and social facts intermingle, creating a complex web through which members of the public must interpret the world. Journalists must therefore become more expert both in gathering information from the pulsating digital networks that now compose our public sphere and in understanding the technical dynamics that characterize the flow of information in networks. This chapter takes a deep dive into both the history and science of networks.

With society's "network turn" has also come a whole apparatus of understanding. From web analytics to network science, practitioners and researchers have developed many new data tools and theories. There is a revolution afoot. One might be faintly reminded of the parallel revolution in economics and finance that took place over the past few decades, as behavioral psychology and the insights of seminal theorists such as Herbert Simon, Daniel Kahneman, and Amos Tversky changed the way the economics discipline thought about the world. Behavioral insights fundamentally disrupted classical models of economics. By harvesting insights from parallel social sciences, economics was able to grapple with the often "predictably irrational" behaviors of human beings—and to grasp reality with greater subtlety and complexity.

For media and communications, such an intellectual revolution is necessary to grapple with a new reality. Increasingly, to understand media effects we must look at networks. *Broadcast diffusion*, a one-to-many hierarchical pattern, characterized media of the twentieth century. Now we must examine a many-to-many pattern—a networked pattern. This network turn in understanding has already begun. Researchers note that social network theory has become a gold mine in the social sciences.[1]

One of the truly astonishing findings by researchers over the past two decades has been that the structure of networks across multiple domains—from cell biology to protein interactions to the World Wide Web and social media—have similar features, or a common *typology*.[2] As we will discuss, some of the typical features of these structures help make the behavior of networks more predictable, both in terms of the flow of information and the overall growth patterns of networks. Knowing these typologies can facilitate the work of fostering networks of recognition, as well as understanding the patterns of social facts.

Certain fairly predictable patterns characterize the development of networks. Many are "undemocratic" in character, so to speak, in terms of structure. This includes the near-universal phenomenon of what is called *preferential attachment*, wherein nodes that have even small advantages or greater attention at the beginning soon become enormous hubs and achieve centrality within the network. This "rich get richer" pattern—also called the "Matthew effect" by the sociologist Robert Merton (in a reference to the New Testament parable of the talents)—does not characterize every network, of course.[3] But the structure of networks rarely looks random, with edges between nodes distributed fairly smoothly and predictably. Networks tend to look like the World Wide Web itself, with a few behemoth sites, such as Amazon or Facebook, and many hundreds of millions of sites that see few links or visitors. This is the same pattern that characterizes content shared online: a few big successes alongside many pieces of content hardly viewed at all.

For journalists, it is well worth getting to know some of the typical patterns and typologies observed in digital networks. The Pew Research Center's Lee Rainie, in collaboration with social scientists Itai Himelboim, Marc Smith, and Ben Shneiderman, has usefully articulated six archetypal patterns, or typologies, that constitute most discussions on Twitter:[4]

1. *Polarized crowds*, which are sharply divided networks and typically manifest themselves in political controversies
2. *In-groups or tight crowds*, which are unified groups centered on professional topics, hobby groups, and the like
3. *Brand-oriented clusters*, or fragmented networks in which many relatively isolated persons talk in small circles about popular products, celebrities, or popular culture

4. *Community clusters*, in which a single topic may be discussed from the varying points of view of different groups

5. *Broadcast networks*, in which a central node such as a large media organization sends out a message or story that then cascades outward

6. *Support networks*, which show large-scale interaction between service providers such as businesses or the government and members of the general public who have concerns, complaints, or feedback

Each of these six typologies can be distinguished from one another by the relative density of the network, or how clustered the nodes are, the degree of division and fragmentation, and the direction of the edges—whether there is conversation going back and forth. Knowing these archetypes and being able to spot and analyze them begins to demystify the world of networks (see figure 4.1).

Yet there is a further reason that journalists should be keen to understand the typologies of networks at least conceptually. Journalists are almost always looking for trends. There is a spate of shootings; people are getting sick from poor sanitation in restaurants; real estate values are booming. Apparent trends are almost always the "news hooks" that justify coverage.

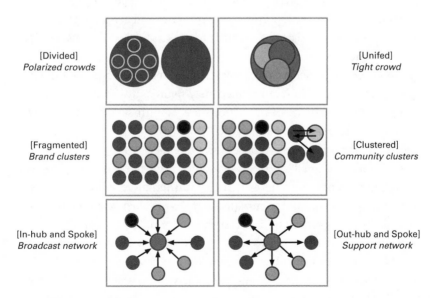

Figure 4.1
Six kinds of Twitter social media networks.
Source: Pew Research Center/Marc Smith.

But in the online world, appearances are not always reality. Many "trending topics" are manufactured. Through marketing campaigns and manipulation, small groups can make their issues and their arguments appear much larger and better supported. Using *bots*, computer programs that spawn automated accounts, misinformation can be promoted in clever and often hidden ways.

Journalists cannot detect all forms of manipulation, but they can get better at not being fooled. Marc Smith, one of the authors of the Pew study and director of the Social Media Research Foundation, says that journalists could benefit from learning how to analyze networks.[5] The visualization software he works with, NodeXL, is one popular application for network visualization. He sees this type of network visualization and analysis as equivalent to taking a "picture" of a given scene. Social network analysis (SNA) tools can help journalists deal with the scale of social media, where frequently it would take reading thousands or millions of posts to discern any pattern or detect manufactured trends. He points out that when a journalist is told to report a particular story, "most of the time that story has a social media component." He adds that "when they go to do the reporting on the social media component, they should bring a digital camera, a virtual camera. That's certainly going to make their reporting easier." Such practices can also make reporting more powerful and accurate, giving journalists the necessary data from which to make sound selections, choices, and news judgments.

Moreover, social network analysis can help reporters figure out the network of influence. "They can immediately figure out who the factions are and who the leaders of those factions are," Smith says, "and what web resources those factions are linking to." Popularity—who gets the most followers and retweets, for example—is just one metric of centrality in a network. Sometimes this kind of evaluation of aggregate data can hide who is truly important. Smith notes that two further, more subtle measures of influence may be useful to journalists: what are called *betweenness centrality* and *eigenvector centrality*. The former is a formula for figuring out who connects the nodes in a network.[6] It may not be the most popular person, but it is someone who bridges different factions. Such persons are the slender threads that sometimes connect to otherwise separate groups. As mentioned, this bridging role is a hugely important one in network science, as bridges allow information to flow and travel.

Then there is eigenvector centrality. For an individual, this is a measure of the influence of his or her friends. Are the nodes around you well connected? Do your friends have many friends? Google and other Internet companies have used this formula to rank the importance and influence of given web pages and pieces of content. It is another tool for figuring out who might be consequential in a given area, a key question for journalists as they investigate or report on any policy area or complex community and attempt to knit networks of recognition.

Twitter: A Cautionary Note for Media

Journalists who participate actively in social media and get information there are participating in a sea change for journalism, which has traditionally kept its distance from sources. Twitter and Facebook now see robust engagement from journalists of all kinds. This can be a net positive, as journalists look to engage audiences and develop sources in wider and less closed ways. Yet as this transformation continues to unfold, it is worth asking hard questions about ethical lines and issues relating to impartiality. Editors continue to monitor how these dynamics are unfolding, and they have not always been happy with the results. For example, the *New York Times* in October 2017 had to make a very public reassertion of rules for reporters on social media, prompted by concerns by Executive Editor Dean Baquet that journalists were stepping over the line in offering opinions on Twitter. "If our journalists are perceived as biased or if they engage in editorializing on social media, that can undercut the credibility of the entire newsroom," the new guidelines state.[7]

My Northeastern University colleagues and I have also been asking some questions about journalistic participation on Twitter. The following analysis is based on a 2017 computational social science study that I conducted with my Network Science Institute colleagues Kenny Joseph, Thalita Dias Coleman, and David Lazer.[8] If there is a main takeaway from this research work, it is a warning that fits in with the previous discussions of how the architectures of networks can leave some things in and some things out. Journalists themselves construct their social universe online, and escaping the filter bubble that they may create can be difficult. Media producers must become more aware of the ways their online networks may inadvertently introduce bias into what reporters think is important (agenda setting, or, in its contemporary form, attention guiding) and how they frame issues.

The designated place online where journalists tend to hang out these days is the Twitter platform. Journalists have found it a kind of water cooler for the digital age, a place that sources, audiences, and peers like to hang out to discuss the issues of the day. The available survey data bear out the anecdotal impression that journalists are spending a lot of time on Twitter—and that they consider it very important for their work. More than half of journalists say they participate in microblogging, and more than half say they find sources on social media.[9] What is unclear, however, is how all this time on social media influences their work, an issue that is difficult to pin down. Research in the journalism studies area suggests that social media signals can be powerful in terms of influencing the behavior of news organizations, editors, and reporters.[10] Further, the sociological literature on social networks and influence consistently notes that our friends and acquaintances can have profound effects on our emotions, behavior, and beliefs, and large-scale studies have established a connection between online signals and offline/off-platform behavior.[11]

We set out to study how the social networks constructed by journalists on Twitter might be correlated with the political shadings of the stories they publish. This required that we select a group of journalists participating on the platform, analyze the persons they followed and code them for liberal-conservative ideological leaning, and then apply another coding system to the ideology of the stories that they published. In other words, we constructed measures of journalists' ideology as represented by who they follow on Twitter and the news articles they write, and then we looked at the strength of correlation using a regression analysis. Of course, these measures can only serve as proxies for the actual ideological leaning of journalists and their stories. The analysis was based on about five hundred thousand news articles produced by roughly one thousand journalists at twenty-five different news outlets.

Using the accounts of members of Congress, as well as a group of twelve thousand accounts of people who can be verified as politically active (both through voter registration rolls and by accounts of officeholders they follow), we trained a model to help us produce a score for the ideology of a given journalist's Twitter network. Immediately, we could see some validation of this model. Journalists with the most heavily right-leaning followerships were at traditionally right-leaning outlets such as the *Washington Times*, Breitbart, and *The Hill*. Similarly, journalists following the most left-leaning accounts tended to be from left-leaning outlets. However, there were interesting exceptions. For example, among the journalists following the most right-leaning accounts were a few stray journalists at *Politico*, the *New York Times*, and the *Washington Post*. Why? It turns out that some of these individuals have beats such as covering the Trump administration or the Republican-led Congress. This points to a weakness in the method for special cases in which a journalist's very focused beat may require him or her to quote specific kinds of phrases often voiced by office holders and policymakers. However, what this potential weakness doesn't invalidate is the general pattern, pointing to the potential of a journalistic filter bubble effect (see figure 4.2).

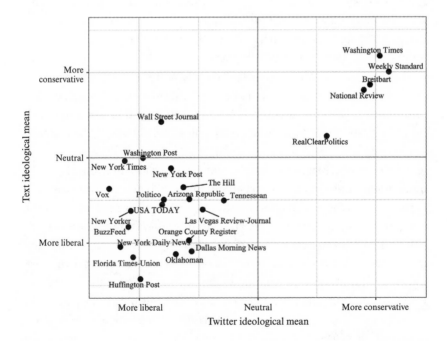

Figure 4.2
Exploring the ideological nature of journalists' social networks on Twitter and associations with news story content.
Source: John Wihbey, Thalita Coleman, Kenneth Joseph, and David Lazer.

In terms of scoring the ideological leanings of the stories, we leveraged the corpus of 150,000 congressional public statements from Vote Smart. After scoring all terms, we extracted the top one hundred most left-leaning and right-leaning terms. Examples of the terms include *lgbt*, *equal pay*, and *voting rights act* for left-leaning persons, and *bureaucrats*, *illegal immigrants*, and *sponsor of terrorism* for right-leaning persons. After further reducing these down to fifty for each political side, we scored journalists' aggregate stories for ideological leaning based on the usage of such terms. It is, admittedly, a limited measure, but our testing it against the collective stories of various outlets—for example, Huffington Post and Vox on the left, Breitbart and *National Review* on the right—suggested that the method had validity.

Overall, we found a clear correlation between the ideology of a journalist's Twitter network and the ideology of his or her writing.[12] This is just a correlation—it does not imply causality, necessarily—but what it provides is a reasonably strong basis to argue that journalists may need to become more aware of their information and sourcing diet, particularly as journalists spend more time and energy on platforms such as Twitter. Of course, journalists are fiercely independent, and a lot of the job is trying to wade through confusing and contradictory claims. As a breed, journalists like to think they can resist the temptation to take information at face value. Most of the time this is true; it is not as if journalists ingesting lots of information of a partisan kind on social media automatically makes them biased. There's no mechanical process at work.

This kind of research on social influence in online networks is all the more important in an era when claims of media bias are rampant. The toxic partisan atmosphere of recent years should not deter journalists from evaluating and exploring their own habits, routines, and information-seeking practices with greater critical scrutiny. Indeed, in my view, we should have the courage to be better and to see our online networks for what they are: a powerful set of signals, cues, and nudges that may frame the way we see the world. Assessing this candidly and more scientifically can help serve the public, which has a greater need for unbiased, professional news now than perhaps ever before. The architecture of networks, and the communities we create within them, can be at once empowering and limiting. Knowing both capabilities and parameters is essential for intelligent and ethical news production in the age of networks.

Scale-Free Networks

As discussed previously, there has been an ongoing revolution in our understanding of many social and scientific domains based on developments in and applications of network-oriented research. The number of persons dedicating careers to making new insights has grown enormously, even though the discipline known as *network science* dates back just over a decade, to the 2005 publication of a National Research Council report that solidified the term and field.[13] Many now refer to this new field as the science of the twenty-first century. Knowing something about the field's development and the contours of some of its major insights can be highly useful to the practice of networked journalism.

A truly foundational insight in this regard—one that helped launch the discipline proper—was made by Albert-László Barabási, a Hungarian-born physicist who has become one of the world's most-cited scientists. Studying the structure of links in the World Wide Web almost two decades ago, Barabási, a director at Northeastern's Network Science Institute, came to hypothesize that most large-scale complex networks are what he called "scale-free," meaning that links are distributed according to a "power law."[14] Put simply, this means that most nodes—or websites, in the case that Barabási initially studied—have very few inbound links to them, whereas just a few nodes have an enormous number of links. There are big winners, in other words, across most large networks, and lots of losers. One might assume that there would be an enormous number of nodes with a kind of "middle" number of links, but it turns out not to be the case: the properties of large-scale networks are such that the distribution of links does not conform to a regular curve of any sort. Hence, they are "scale-free."

Why does this matter? There are many reasons, with many discipline-specific nuances. Yet as Barabási writes, "Notwithstanding the amazing differences in form, size, nature, age and scope of real networks, most networks are driven by common organizing principles."[15] There seems to be a central, universal tendency. This principle helps frame the evolving landscape of digital media and information.

The laws of large-scale networks have implications for news. "With the fragmentation of media, hubs become even more important," Barabási says.[16] "With the democratization of the process, the larger the network, the larger the hubs." A couple of decades ago, as the web began to expand exponentially, there was an expectation that this vast new sea of information nodes would in some ways dilute the power and reach of the traditional news media players. This has happened to some degree, but not for the very biggest players necessarily. In fact, the opposite has happened in many ways. For example, the *New York Times* and the *Washington Post* are now seeing a combined 150 million unique visitors monthly—representing some large fraction of the global, English-speaking population. Other big news outlets have also vastly expanded their ability to reach and influence audiences. As the market increased, the biggest hubs increased significantly in size.

Further, as the market increased, the smaller players have receded in relative impact and reach, victims of a mathematically predictable "rich get richer" or power law phenomenon. Globalization, Barabási explains, undercuts the model of news being served locally. The consequence of the overall information and news network growing massively is the relative diminishment of local news in its ability to find a strong role (the inability to attract advertising is evidence of this, as advertisers flock to big hubs such as Google and Facebook or other large media brands). In addition, the democratization of information has had surprising effects on large traditional publishers of many kinds, even as their business model has been shaky. Barabási notes that the top scientific journals, for example, were expected to diminish in importance with the rise of the web; yet what has happened is that there is even greater value and prestige attached to them. The same might be said of the top news brands. "There is value in gatekeeping when networks increase in size," he says. "With extreme production, quality control becomes a premium." Gatekeeping, almost perversely, finds an even stronger role in many domains as networks expand.

Small Worlds

The discipline of network science combines several threads that developed roughly in parallel in mathematics and social science. Paul Erdős and his collaborator Alfréd Rényi published a paper in 1959 that furnished the early groundwork for the discipline of studying networks, establishing the basis for *graph theory*, the mathematical study of relations between objects.[17] Yet as that mathematical tradition was beginning, a somewhat separate intellectual genealogy was also growing. That latter network of scholars extended back to European-born social theorists such as the late-nineteenth/early twentieth-century figure Georg Simmel—acknowledged as the first to think about human social relations in explicitly network, or "geometric," terms—and Jacob Moreno, an eccentric but highly creative psychologist who in the 1930s drew the first sociograms, maps of human relations using nodes and edges. As the world was becoming more urbanized in the early twentieth century, more theoretical and empirical work began to take place in the context of cities, and the meaning and consequences of increased incidental contacts with others became of interest among scholars. This would eventually lead to the creation of the discipline of social network analysis.

From these early origins, interest in social networks and "structural" understandings of human relations began to be codified and extended by the likes of Talcott Parsons, Harrison White, Robert Merton, and Paul Lazarsfeld from the 1940s on. They generated their own branching intellectual network, as their many graduate students would go on to become influential social network theorists themselves.[18] Lazarsfeld produced pioneering work in mass communications research and statistical analysis of media and social influence. His landmark book with Elihu Katz, *Personal Influence*, helped cement an idea that has continued to define the study of the spread of ideas through media. Called the *two-step flow of communication*, the main idea is that much of social learning about the world happens through opinion leaders who mediate between channels of mass communication and individual citizens.[19] Citizens do not directly understand media messages in the way a hypodermic needle injects a person. That focus on the behavior and importance of opinion leaders—also called *influentials* or *connectors*—has become a mainstay and pillar of social media marketing and all manner of online media strategies.

The branches of the winding intellectual backstory of research in social networks lead in many directions, but certainly one of the central contemporary nodes is Duncan Watts, now at Microsoft Research. Watts remains known for his first research on networks in the mid-1990s, performed as a graduate student at Cornell in the Department of Theoretical and Applied Mechanics, under his graduate adviser Steven Strogatz. A sometime rock climber, Watts decided to embark on observational research that required him to climb trees to study the communication patterns of a certain species of cricket. After working with an entomologist and the necessary climbing and capturing of the crickets, Watts brought the insects to a lab, where he would wait for them to communicate. There was a lot of time to think: "Some nights they wouldn't chirp at all."[20] The way that their chirps managed to become collectively ordered was, Watts thought, potentially a way of looking at the phenomenon of synchronization or dynamical systems theory in a biological context.

At the same time, Watts and Strogatz had been thinking deeply about other, related research literatures that explored connected systems, including Milgram's "small world" experiment using postcards through the mail, which (as will be discussed) produced the original idea of six degrees of separation across a population. "While I'm sitting there waiting for the thing to chirp," Watts recalls, "I had been thinking about all of the literature I had been reading on synchronization."[21] He then began connecting the small-world idea with dynamical systems theory. Watts and Strogatz would eventually run computer simulations to illustrate that vastly disparate systems—the biological system of a roundworm, the power grid, the collaborations of Hollywood actors—have a common network architecture. In essence, all these various systems show high degrees of local clustering, with lots of nodes connected by dense edges in regions of a network. As Milgram's experiment suggested, there were short global path lengths—ways of getting nearly anywhere in the network through just a few hops, or degrees of separation.

The empirical work with the crickets would never be finished directly, but the essential ideas had been formed in those late-night hours of lab work. The article that Watts and Strogatz would publish in *Nature* in 1998, "Collective Dynamics of Small-World Networks," would be cited tens of thousands of times over the next two decades—one of the most important works produced in a generation.[22] Others would come along to make further insights in the same domain, but the article heralded the new revolution—what would eventually become network science. "I had this feeling it was a paradigm shifting thing," Watts says. "The first time I had pins and needles and shivers. It was spooky. I thought, 'I've discovered something about the world.'"

Watts would be deeply influenced during his time at Columbia University by Harrison White, whose own earlier research in social networks would exemplify what is sometimes called in the field the *Harvard revolution*, a school of sociological thought that mapped a new scientific approach toward studying human relations and networks. Watts would also have contact with Merton at Columbia, and the relationship proved crucial.

Among the other early figures studying social influence and social group behavior, Merton stands as a giant in sociology, his work helping to define the discipline across many subdomains. In certain ways, we might identify both him and White as essential nodes in this early intellectual space. And from them there would be a handoff or baton passing of sorts to Watts, connecting the older sociological tradition to our present age of media and digital networks.

Bridging Ties: Sociology to BuzzFeed

It was not long after Watts's formative years at Columbia that he would come into contact with and befriend media experimentalist, entrepreneur, and fellow wunderkind Jonah Peretti, a cofounder of the Huffington Post and now known universally in the media world as the founder and architect of BuzzFeed, one of the most successful news media startups in the social web era. In 2007, Peretti and Watts coauthored an influential article for *Harvard Business Review* on viral marketing, arguing that the way to see peer-to-peer network effects around content spread was through subtle statistical patterns. "As we pointed out in that article, things don't have to go viral in the traditional, mathematical sense of the reproductive rate exceeding 1, and showing exponential growth," Watts notes. If the reproductive rate is only half, or 0.5, which appears low, content publishers can still double the number of people reached over time, provided that multiple generations of sharing occur. "We realized this when we were looking at a bunch of campaigns that Jonah had run, long before he started BuzzFeed," Watts says. "And we were trying to find evidence of viral spreading. We kept being disappointed, as we were seeing numbers of reproduction less than 1."[23]

Peretti recalls that those early days of Internet experimentation were influenced by more formal social network analysis and theory, as well as diverse influences from the wide-open ethos of the 1990s and early 2000s web era. "Duncan's work and conversations with Duncan were influential as well as earlier work by social scientists like Stanley Milgram and Mark Granovetter," Peretti says. "I was also influenced by early web culture, blogging, message board culture, del.icio.us, early YouTube, early Twitter, and creators like Ze Frank who now runs BuzzFeed's Entertainment Group."[24]

What Watts's insights ultimately pointed to is that going viral is fundamentally unpredictable, an idea that has seen continued refinement and validation. Only about half of the factors that help predict a given viral cascade can be predicted.[25] Too much depends on the fickle and unknowable receptivity and structure of the network, properties outside the control of media producers. Targeted experimentation within the network therefore is the only sensible strategy to generate the best chance of marrying content type and presentation with the receptivity of social networks. Content must be "tuned" in a sense to the pitch of the digital environment. Data-driven testing is not a silver bullet, but it's absolutely necessary to maximize reach.

Founded in 2006—in the same incredibly fertile period in which Facebook, Twitter, and YouTube were launched—BuzzFeed was pitched to multiple venture capitalists as a pioneering attempt to capitalize on the virality and quasi-virality of online content, based on a philosophy of experimentation. "I remember in the early days when Jonah was going around pitching BuzzFeed to potential investors, the [venture capitalists] just couldn't wrap their heads around it," Watts recalls. "[He] was saying 'no one knows how to the find the influencers, so instead we're going to trigger social contagion in a probabilistic way, playing the odds and learning what works from data.'"[26]

Time and again, across numerous empirical research projects, Watts has hammered away at the theme that virality is both rare and unpredictable, sometimes attacking Gladwell for a credulous belief in the power and reliability of influencers. (Watts's 2003 book *Six Degrees* glancingly critiqued Gladwell's 2000 book *The Tipping Point*, but a few years later Watts took the "influencer" model head on.)[27] The idea of studying virality empirically is something Peretti has taken to heart—and leveraged with massive success—and BuzzFeed's digital strategy and philosophy are predicated on this idea in many ways.

Indeed, it is only now that everyone across the web, especially in news, is "getting" the strategy, adopting and appropriating many of the same experimental and data-driven techniques and philosophies. Technology and legal scholar Tim Wu notes that Peretti's early, original theories, "'particularly about the necessity of stimulating 'pleasure in the social process of passing,'" would eventually be validated by their near-universal adoption. The techniques that Peretti developed across numerous provocative and often humorous experiments in viral media prior to the founding of BuzzFeed, and then subsequently formalized and accelerated with BuzzFeed, have become ubiquitous across digital media: "Collectively BuzzFeed and its rivals—Mashable, Upworthy, and in time parts of the mainstream media—began to crack the code; eventually they could consistently make content go viral."[28]

BuzzFeed has always been a mixture of data-driven exploitation and human-intuitive exploration, leveraging social contagion without assuming that it can be predicted (i.e., the original Peretti-Watts thesis). It pioneered the idea of what Peretti called, in a derivation from Watts, "viral lift"—a measurement of the sharing rate related to any given piece of published content. This idea of measuring over- and underperformance of digital media content is now commonplace for social media desks across most news outlets, aided by applications such as Chartbeat and NewsWhip. Peretti notes that the company is obsessed with the metric of peer-to-peer sharing, as he argues it generates the best window into the collective receptivity—relevance and interests in real time—of the network and what precisely is "creating a social connection between people":

> It is why we obsess about "share statements," or what people say when they share our content. It explains why we carefully study the exchange of value that occurs when sharing happens: What is the value for the person who shares? What is the value for the person who receives? And what does the activity mean for the bond they share with each other? Understanding the inherently social nature of media is one of the biggest "digital advantages." But despite the rise of social media and the growth of BuzzFeed, it will still take years until most of the industry fully understands.[29]

Because of its use of formulaic "listicles" and its many bits of frothy and frivolous content, BuzzFeed often gets sneers from serious news observers. But as it has both built a substantial news division and expanded into many kinds of media production, the model looks more intriguing from a news sustainability perspective. The company registers tens of billions of content views annually across social media platforms.[30] With its mix of entertainment and news, it has been likened to a kind of twenty-first century television network, serving a broad range of content and letting the frothier fodder subsidize the hard news.[31]

More than a decade beyond that initial *Harvard Business Review* article and intellectual collaboration, Watts and Peretti remain friends, with Watts offering technical and strategic advice on data science projects, such as BuzzFeed's social and online tracking tool Pound. BuzzFeed's strategies and pioneering work in data-driven viralization of media content have influenced countless fascinated imitators and other news outlets, all eager to capitalize on the phenomenon of peer-to-peer network efforts on social platforms.[32]

Changes in Community Ties

Let's imagine a media producer gazing out into the digital world, staring into her screen, typing away, editing video, perhaps coding an application. On the other side, somewhere in the wires that connect to billions of other computing devices, is a "public" that constitutes the relevant audience for the journalist or the set of sources that might help inform or illustrate a story or trend. What might be the actual shape of that public? And how is this wired public any different than publics in past eras of human history?

Instead of thinking about the interconnections among billions of persons, it is worth first thinking about each person as having a network neighborhood. This network neighborhood has changed in certain ways in the digital era, yet it also shows certain continuities even with our predigital ancestors. Cultural norms, geographic remoteness and proximity, and the architectures of social life in different societies all serve to shape the number and quality of our ties or social contacts. However, there is good evidence that humans, on average, are able to sustain a maximum of about 150 relationships of various kinds. The anthropologist Robin Dunbar has argued that there may be natural "cognitive constraints" on trying to maintain networks larger than this.[33]

Social media platforms allow us now to maintain connections with much larger numbers of people—high school classmates, coworkers from prior employment, and so on—with minimal effort. This is an apparent change from the past, when deliberate, relatively costly effort through letters, telegrams, phone calls, and so on needed to be employed to maintain ties. That is not to say that all of these expanded connections are influential or powerful in an individual's life, but there are more of them. Dunbar's more recent research on Facebook users suggests that social media are not typically used to expand social circles; rather, social media may merely "function to slow down the rate of decay" of friendships, which, lacking face-to-face contact, naturally dissolve.[34] Further, the patterns of interaction in online networks "reproduce rather faithfully both the nested structure of the inner layers of offline networks and their typical interaction frequencies." In other words, digital trace data show that most people have roughly the same number of close friends online, with whom we interact frequently, as we do in our physical, offline realities.[35] Data generated through the General Social Survey in the United States—which has carefully tracked changes in American social life for decades now—do not indicate a general increase in the capacity for friendship or the number of close friends (also called *core discussion networks*) because of the rise of social media.[36]

Social scientists have long been dividing individuals' network neighborhoods into what are called *strong* and *weak ties*. In general, strong ties are closer friends and family; weak ties are casual acquaintances, such as friends of friends, and are characterized by glancing or infrequent interactions. Social scientist Mark Granovetter, whose foundational 1973 paper "The Strength of Weak Ties" highlighted this distinction, defined a strong tie as a "combination of the amount of time, the emotional intensity, the intimacy (mutual confiding), and the reciprocal services which characterize the tie."[37] The salient point made by Granovetter is that strong ties typically do not "bridge" across disparate communities. Why? It is almost always the case that our good friends know one another, and the idea that our two good friends would not know one another is generally a "forbidden triad," he says, putting it in the technical language of social network analysis. Our collective network of strong ties is therefore an echo chamber of sorts. For the spread of news and information, weak ties are needed; they are the primary pathways through which novel information travels.

The power and impact of these forms of ties may vary, and there has been decades of research and debate on related research questions. One of the more important contributions in this regard, and one that brings the original insights of Granovetter into the digital age, comes from Facebook's own data scientists, who in 2012 conducted a field experiment in which Facebook randomized how some 250 million users received information signals from friends. The researchers noted that users are "more likely to have the same information sources as their close friends, and that simultaneously, these close friends are more likely to influence subjects."[38] Overall, Granovetter's claims about weak-tie advantages were validated: although strong ties are much more influential and have more impact on user behavior, it seems to be the "more abundant weak ties who are responsible for the propagation of novel information."[39] (Another paper from the Facebook data science team has presented similar findings with respect to the influential nature of strong ties as compared to weak ties.)[40]

Yet the debate remains unsettled, particularly as social media platforms evolve. For example, technologist Ethan Zuckerman has asserted that the online world may be fundamentally restructuring social ties. Granovetter's ideas, Zuckerman writes, "may have been true in 1973 when he wrote the paper, but they are more questionable today." Connectivity has increased, diminishing the role of physical geography. Therefore, according to Zucker-man, "In an age of digitally mediated friendships, it's quite possible—and likely quite common—for strong ties to be bridge ties."[41] This would be consequential and novel in human communities, as it would mean that powerful, behavior-shaping information (that propagated by strong ties) would more readily reach disparate communities (groups of weak ties).

There is some empirical data to support this idea of greater numbers of "bridging" strong ties. "It is commonly the case in people's offline social networks that a friend of a friend is your friend, too," researchers at the Pew Research Internet Project found in a study that sampled social media users. "But on Facebook this is the exception, not the rule. ... The average Facebook user in our sample had a friends list that is sparsely connected."[42] Those sparse connections mean potential links to communities, and news and information, that one otherwise would not find.

I asked Granovetter what he makes of this debate and the state of play of strong-weak tie dynamics in the social-digital era. Interestingly, he remains unconvinced that these strong-weak tie relationship dynamics are chang-ing dramatically in the digital age. "I'm not sure why strong ties would serve this function in the digital world any more than in the 'real' world," he says. "There has been a lot of research on large datasets, mostly cell phone calls rather than social media sites to be sure, and in that research, weak ties continue to play a special role in bridging. ... In fact I would sus-pect that given the reduced cost of maintaining weak ties that results from the virtual world, their role as bridges might even be enhanced."[43]

Obviously, relationships do not always neatly fall into strong and weak categories. Researchers performing social network analysis have studied this problem extensively in recent decades, focusing on issues such as which kinds of digital trace signals best indicate tie strength and how the gen-eration of new ties and relationships might be predicted given existing networks.[44] Some research has suggested that the sweet spot for maximum information diffusion may not be either through weak or strong ties, but through ties of intermediate strength. Interactions with weak ties are too infrequent to reliably pass along information, while strong ties constitute, as discussed, a kind of closure that makes novel information less likely to penetrate or ripple outward. Intermediate ties, then, allow for both suffi-cient frequency and novelty to allow for information diffusion.[45]

Through research in the domains of health and medicine, Nicholas Christakis and James Fowler have been trying to establish some fundamental laws in the offline world regarding ties.[46] Their research has created awareness that friends can be hazardous to our health, so to speak. Because of the viral power of peer effects, the behaviors and conditions of those in our social circles are somewhat contagious: obesity, drinking, smoking, and drug use, as well as depression and loneliness, can all spread to us. These contagious effects appear to ripple across three degrees within networks—friends of friends of friends. Christakis and Fowler call this the "three degrees" rule. They showed the multilink causal chains of health in social networks beginning with their 2007 study in the *New England Journal of Medicine* that analyzed Framingham Heart Study data and showed how weight gain could spread through networks.[47]

How exactly such findings translate into a digital context is unclear. A 2014 experimental study by Facebook showed how users' news feeds could be tweaked to spread both positive and negative emotions.[48] That study explicitly cited and drew on the work of Christakis-Fowler, and it showed how emotions could ripple through networks across several degrees of connection over the Facebook platform. (It also became a public relations nightmare for the company when it was published due to allegations of unethical treatment of human subjects.)

Revolutionary Bursts and Digital Ties

How is the expansion of weak ties through social media affecting the capacity of news and information to spread beyond what sociologists sometimes refer to as *cliques*, or communities that are tightly clustered? On occasion, we continue to see patterns of large-scale activism that simply would not have been conceivable prior to the advent of peer-to-peer communications technologies. These kinds of events—whether the Black Lives Matter movement, the 2017 Women's March in the United States, or the Arab Spring that began in late 2010—certainly have something to do with the expansion of weak ties online.

Social scientists have begun evaluating the apparent sea change in the nature of collective action and activism because of digital technologies, what Lance Bennett and Alexandra Segerberg call "connective action." They note the sharp distinctions between social movements past and present: "Compared to many conventional social movement protests with identifiable membership organizations leading the way under common banners and collective identity frames, these more personalized, digitally mediated collective action formations have frequently been larger; have scaled up more quickly; and have been flexible in tracking moving political targets and bridging different issues."[49]

The speed with which mass movements can appear and grow around, for example, a hashtag is impressive, but it is worth noting that some movements that are organized in these more informal, peer-to-peer, issue-based ways do suffer from a lack of sustainability. Zeynep Tufekçi has noted the paradoxical rigidity of such movements, which suffer from a "tactical freeze" once the initial protest phase is over, as they frequently cannot come to a consensus about tangible policy demands or further waves of action once they face challenges and resistance.[50] Weak ties cannot become strong ties quickly; solidarity, consensus, and priorities take more time than digital organizing sometimes allows for.

What is emerging is a picture of a social world that has changed in important ways but that retains features of the past. Barry Wellman and Lee Rainie have usefully offered the framework of networked individualism, what they call a new "operating system" for social behavior predicated on looser, more fragmented networks.[51] Wellman's long-running sociological work suggests that many of these changes were taking place even prior to the advent of the Internet and that "small, densely knit groups like families, villages, and small organizations have receded in recent generations." Still, information and communication technologies have "powerfully advanced" the shift from small-group networks to "broader personal networks."[52]

We should be careful about interpreting digital evidence that appears to suggest dramatic shifts in human behavior due to the Internet, even when we can see large apparent effects. In a landmark study, Fowler and a team of researchers evaluated Facebook's large-scale experiment on voting day in the United States in November 2010, involving more than sixty million people. As part of a randomized trial, Facebook displayed "Go Vote" messages in some people's news feeds and facilitated people signaling their actions to friends by clicking "I Voted." The researchers estimated that 340,000 additional votes were cast because of the experiment; they also found that people responded most strongly to messages of friends with whom they had a strong offline relationship. Indeed, the influence of close friends was about four times greater.[53] Strong ties, likely nurtured offline, too, still prevailed in producing the most substantial effects. The empirical evidence has continued to suggest a reasonably tight relationship between offline and online worlds, Fowler says: "Everything that we have done so far has suggested that the changes that happen online are changes that could have happened in real life. The online world just facilitates those changes."[54]

Yet in the domain of news and information diffusion, there is a special dimension of digital social networks that seems, on its face, novel and truly distinctive, as compared to communications networks of the past. What continues to tantalize everyone in the media world is virality, which can potentially take a media item from obscurity to global visibility in a matter of minutes. This is known as the *contagion effect*, and it is often misunderstood. We saw virality play a role in the work of Watts and Peretti. Let's take a closer look.

Science of Contagion and Virality

Imagine again the scenario of the media producer gazing out into the digital world, gauging the potential audience for a story—and the potential nodes who might constitute a network of recognition. There are billions of individual neighborhood networks, perhaps about 150 "core" persons per individual, with many digitally enabled weak ties. These weak ties mean that an individual's neighborhood network is likely connected to more disparate neighborhood networks than ever before. In these interconnected networks of networks, certain persons will have a higher "degree"; that is, they will have more edges connecting them. As discussed, in the language of networks this is called *betweenness centrality*. Theoretically, reaching such strategically placed persons might help a piece of content spread faster, although the research on this issue is complicated. There is a long-running research question about how much these well-positioned connectors matter in terms of viral spread versus how much the overall structure of the network and its receptivity to certain messages determine virality.

Each potential audience member will engage with a given story in one of three basic ways: (1) media (broadcast) activation, or the sending of content from larger media entities directly to individuals both online (email, news alerts, direct audience access to websites) and/or through radio, television, print, or other forms of mass distribution; (2) viral (peer-to-peer), which is defined by direct sharing among individuals, typically on social media platforms or through email, text, or messaging applications, without intervention by large media entities; or (3) hybrid (broadcast and peer-to-peer) patterns, which combine elements of traditional and social media. As we will discuss, almost all the media consumed on the web comes through either media activation or hybrid patterns; a very small percentage finds its way to audience members through a pure viral effect.

Going viral by leveraging these networks of weak ties is not easy. Using experimental techniques, Solomon Messing and Sean J. Westwood have shown that for media content, strong ties are key to sharing and directing attention. Tie strength, they note, is "a strong determinant of attention to traditional media items on social networking websites like Facebook." We are less likely to care about what casual acquaintances have to say, even if they are more likely to bring us novel information.

For a piece of media content to go truly "viral" in the technical sense, it must pick up a certain amount of velocity and replicate itself rather quickly. This may be quite different than the usage of "viral" in common parlance, where it may just mean something that was very popular on social media (and that often became popular because it saw attention from large media outlets) or a slow but wide-crawling meme or media item that has a very long tail in terms of attention. Facebook's data scientists have noted that photos shared can continue to cascade slowly over periods of up to four weeks.[55] Based on a review of the literature and their own empirical work, Karine Nahon and Jeff Hemsley define *virality* as "a social information flow process where many people simultaneously forward a specific information item, over a short period of time, within their social networks, and where the message spreads beyond their own social networks to different, often distant networks, resulting in a sharp acceleration in the number of people who are exposed to the message."[56]

This sharp degree of spreading, which in the language of web analytics is roughly synonymous with the term *performance*, is the key component here. It is important, too, for more than academic definitional reasons. Some of the leading researchers affiliated with the Facebook data science team have found that sharp, early viral spread is also associated with "burstiness" later on; in other words, content may be more likely to see further secondary and tertiary accelerations if it sees initial virality. Research has found that the specific character of the initial burst of sharing is quite important as well: "The virality, or appeal of a cascade plays a role in recurrence: cascades whose initial bursts are long-lasting, moderately popular, and moderately diverse are most likely to recur."[57]

There is something special about content that sees intensive initial sharing, a phenomenon that likely speaks to the receptivity, or ambient readiness, of the audience, like a wave that is being surfed. For example, a photo or blog post that speaks to social justice issues might see initial burstiness if public discourse has recently focused on related issues. The content itself is not only compelling but well timed and tuned to a primed network ready to engage with it. This flammable, ambient readiness and fertile receptivity embedded in the network structure help explain, at a scientific level, the phenomenon of networks of recognition. Through such networks, information is passed from node to node, creating a fast phase transition across a given network.

The largest study done on online virality comes from a combined team at Stanford and Microsoft Research, which analyzed an entire year's worth of Twitter activity: more than one billion links shared in the categories of news, images, videos, and petitions. Roughly one in every three thousand links triggered a "large event," reaching at least one hundred persons. Yet large-scale viral events, defined by multiple generations of peer-to-peer sharing and a reach of several thousand adoptions at least, occurred on the order of about one in one million links shared.[58] "These are very, very rare events," lead author Sharad Goel says of the study's implications. "Almost always people are getting information directly from popular sources, and they don't pass it along."[59]

What does virality actually look within the network? The truth is that the metaphor of disease, of a literal virus, is not really appropriate. In a biological network, a virus has a chance of spreading to every individual without reference to any external factors; one is either immune or not. By contrast, because humans are social animals and become increasingly prone to doing certain things as their community moves in a given direction, virality has a direct relationship to the neighborhood network of a person. Such phenomena are called *complex contagions*, for which reinforcement will be required to overcome a certain threshold before someone engages with a given piece of content.[60] This might mean, for example, three or more friends sharing or engaging with a given news story on Facebook. The threshold is the number past which behavioral change or adoption of an idea becomes likely.

It is a paradox of sorts that highly connected networks—dense networks of links all following one another—are actually less likely to produce influential ideas, memes, and information cascades. They are not good at producing global cascades that bounce from community to community. When communities are highly connected, they tend to trap information, because it takes a lot of reinforcement from multiple persons to generate adoption and attention. No one person has a monopoly on influence in such dense networks. Most hashtags, in fact, become trapped in highly connected communities because of these dynamics.[61]

Yet it is also true that networks that are not connected sufficiently also have a low chance of diffusing viral content, because there are fewer ways for information to jump to other communities. There is a trade-off, then, between networks that are densely connected and those that are sparsely connected. On rare occasions, enough thresholds will be crossed in just the right place—in a place that is highly connected within a network—that a cascade will occur. But again and again, researchers have found these viral events to be rare in social systems, because the vulnerable area in which a cluster or chain of interconnected persons can influence others is nearly impossible to pinpoint in advance.[62] Large-scale networks of recognition are difficult to achieve, but knowing the regular laws that govern networks can help.

5 Bias in Network Architectures and Platforms

Information networks always have an initial architecture, or at least some set of parameters. Their design is important: it enables and prohibits, slows downs or speeds up, certain kinds of activities. Protocols internal to the network might dictate who is in and who is excluded. Awareness of these architectures and features is useful on a variety of levels for journalists and others involved in communications: it can help facilitate community engagement and better sourcing while also overcoming biases that are inherent to networks.

Gaining greater insight into the constructed nature of modern digital networks, such as social media platforms, is imperative for news media. Fostering networks of recognition requires an awareness of how ICTs can drive human communities apart and the structural barriers to be overcome. Algorithms and web applications are increasingly governing what the public—and what journalists—see and value. Journalists must retain their core autonomy, even as the idea of press autonomy must adapt to new circumstances within an expanding infrastructure. As Mike Ananny notes, "Press freedom is predicated on the power to realize publics, and … for the networked press, this power resides in its configuration of sociotechnical relationships."[1] These relationships are complex and constantly varying, but there are some durable principles relating to network architectures that we can outline.

The discussion that follows proceeds through a number of domains and applications across the modern history of telecommunications technology. It uses several moments of innovation as its central case studies, including the wiring of the telegraph across the North American continent in the nineteenth century, as well as the fascinating early network-engineering efforts of Sergey Brin and Larry Page, the founders of Google, and Mark

Zuckerberg, the founder of Facebook, in the late twentieth and early twenty-first centuries. Exploring the story of the telegraph makes more concrete the idea that networks are "engineered" through design choices, with consequences for how information flows. The network architectures created by web search and social media pioneers Brin, Page, and Zuckerberg now arguably are as important in our time as the networked physical infrastructure that knits together the modern world, with its highways and bridges, telephone poles and airports. It is worth understanding them better.

One obvious way in which digital information might be influenced or indeed curbed is through web filtering by a government—something that takes place in many countries, with China being the most obvious practitioner. Suppression of content and online monitoring by government officers and their networks can significantly shape what kind of information spreads and to whom. But that is only part of the picture. Other key choices are inherent in decisions both at the network level and in the applications that are designed to navigate the web. Certain constraints can influence what kinds of information are shared and how they flow. Some are basic: think of the (original) 140-character limit on Twitter or the various limits on video length on social platforms. But also consider the algorithms that select items to appear in someone's Facebook News Feed or the search algorithms that dictate which results are seen on the first page of a Google query.

What social scientists sometimes call *affordances*, or the structures of features in applications that enable certain kinds of use and behavior, can influence the overall strength or weakness of a network.[2] In a predigital era, Harold Innis called this the "bias of communication," whereas legal scholar Larry Lessig has put a modern spin on it in asserting that "code is law," meaning that technical parameters embedded in HTML, CSS, JavaScript, Python, SQL, C++, and more operate functionally to enable or prohibit human behaviors (e.g., sharing, buying, remixing, streaming, communicating) much in the way that black-letter law has in the traditional physical world.[3]

Why do affordances in networks really matter? Let's consider a couple of examples. In April 2015, Twitter introduced the ability for users to "quote" other messages and then comment on them. This became known roughly as the practice of "quote retweeting" or "quote RTs." This is a relatively subtle change on the microblogging platform, which except for its character constraints (140 characters for a long time, but now 280) is a relatively wide-open and uninhibited platform for communication. Researchers have found, however, that just this change in affordances, allowing quote RTs, improved the quality of political discourse, at least in the initial stages of the rollout of the platform tweak. The change facilitated "increased political discourse and its diffusion"—fostering more deliberation on public affairs topics—and enabled a "more civilized form of communication where people discuss and agree with each other, with far fewer insults being observed."[4] What is important to consider here is that because some communications platforms have such a vast scale—Facebook has two billion users, Google has surpassed two trillion searches a year—even small changes in the architecture have major consequences for human behavior.[5]

Take another example from the world of journalism: the comment threads that have traditionally accompanied news articles. Commenting systems often allow users to evaluate and "vote" on whether they like individual messages from readers. What are the consequences of such systems of crowdsourced opinion? Researchers have found that negative feedback has a significantly larger impact than positive feedback. This begins to answer a perpetual question for online news editors: "Why did the comment thread go so sour?" As anyone who has looked at comment threads appended to news articles knows, many quickly deteriorate into personal attacks and bizarre digressions from the main point of the article; misinformation abounds, and because news organizations lack the resources to monitor comment threads, much of the misinformation goes unchecked.[6] This actually hurts news, because viewers may filter their understanding and memory of articles through the comments. Comments can distort readers' ability to grasp facts in the article.[7] Indeed, a group of Stanford researchers has found that "negative feedback worsens the quality of future interactions in the community as punished users post more frequently."[8] The upshot of this is that commenting systems should find ways to avoid allowing for overly negative feedback (while still allowing for positive evaluations). Commenters frequently interpret negative votes as a form of personal punishment and respond with anger and lower-quality comments, fueling a vicious cycle. This is again a story illustrating that small tweaks in networked communications architecture can matter enormously.

There is a constant interplay and tension between the natural tendencies of networks—the universal shapes that networks typically evolve toward—and the shaping forces and human-engineered constraints and pressures that can guide, influence, reshape, or even outright skew or bias information networks in one direction or another. Communications networks are always constructed. They are a product of human intention, even if their eventual uses take on forms unanticipated by their founding engineers. To get a clearer picture of the constructed nature of networks, let's consider a few of the most important examples from the past two centuries—the original electrical network, the telegraph; and its great-great-grandchildren, Facebook and Google, which have come to define the contemporary communications and information industries.

The Telegraph: Wiring a Continent

Networking the land of Silicon Valley—as it would be known generations later—with the "civilized world" and the Eastern United States was no easy feat. Building such a communications network to the Bay Area in 1861 involved complex government contracts, appropriated by Congress, and coordination with both the Western Union Company and the California State Telegraphy Company, which would string wires toward one another and meet on the east side of Main Street in the telegraph office of Great Salt Lake City. The telegraph line would stretch from Omaha to Salt Lake City and Carson City; San Francisco would wire back in that direction.

This new telecommunications network—this project of wiring a continent—came as a result of an enormous task of engineering and brute force labor, as well as a bit of cross-cultural diplomacy with both the Mormons (from whom much of the pole timber came for the Southwestern route) and Head Chief Sho-kup of the Shosones, who had concerns about the intention of the white men so eager to put up mysterious poles and wires over their territories.[9] To string wire across the forbidding Sierra Nevada, 228 oxen, 50 men, 26 wagons, and sufficient horses were required. Through hot deserts and over snowy peaks, a band of engineers, mountaineers, and Native Americans would stand up between three and eight miles of poles a day. The first message from California back to Salt Lake was sent at 5:13 p.m. on October 24, 1861. Brigham Young himself would reply in kind to the operator in San Francisco. At that instant, a coast-to-coast message became a reality.

In 1881, James Gamble, who was involved in those early, historic construction efforts, reminisced in *The Californian* magazine about this great communications networking project—and the day the East and West finally met. "In that moment California was brought within the circle of the sisterhood of States," he wrote. "No longer as one beyond the pale of civilization, but, with renewed assurances of peace and prosperity, she was linked in electrical bonds to the great national family union." The vaunted Pony Express, symbol of the communications structure of the Old West, had begun to be supplanted by a new technology.

It might seem inevitable from there, in the early 1860s, that the individual human capacity to overcome distance and stay connected would be radically increased, knitting the American people into a kind of proto-Facebook by dint of the telegraph wires. Certainly the telegraph played a crucial role in the American Civil War, and the financial system soon found all sorts of uses for it. Yet for many decades hence, the telegraph did not serve as a personal communications tool for citizens. It would take the mass adoption of telephony in the early 1900s to accomplish that goal more substantially. As the historian David Hochfelder has pointed out, most people did not encounter the telegraph as a communications tool, but rather as a mechanism for public information from other intermediary sources.[10] Telegrams were expensive. The monopoly business of Western Union ensured that the costs did not come down, despite the efforts of reformers who wanted to replicate the cheap postage movement of the 1840s and 1850s, which had in effect democratized the US mail system.

In fact, the telegraph was very much the opposite of a democratic tool that allowed for broadly shared connectivity. Early usage of the continental systems of telegraphy was almost entirely by businesses, government, or news companies such as the Associated Press. Certainly this first electrical telecommunications technology revolutionized society in all sorts of ways. Yet the telegraph "revolution was not instantaneous, and it affected Americans differently according to class, region, and other demographic characteristics," Hochfelder notes.[11] The very Native Americans whose lands were used to convey the wires to the Pacific were not brought into the world of these new technologies, whose advent heralded yet another step in the nation's taming of the West and the building of empire. Who is in and who is out in the structure of a network remain vital questions even today.

It was during this historical period that the very term *network*, often hyphenated archaically as *net-work*, came into broad circulation to refer to communications. The word dated back to the 1500s in English, and according to the Oxford English Dictionary it had a distinguished history, showing up in early translations of the Bible, as well as in verse of the likes of Coleridge ("The arterial or nerve-like net-work of property") and Shelley ("The woven leaves / Make net-work of the dark blue light of day.")[12] With the advent of the telegraph, observers grasped at metaphors to describe these iron lines that were everywhere being spun out overhead, using the language of spiders and webs, mazes and labyrinths.[13] An 1873 report in *Harper's New Monthly Magazine* noted the almost mystical nature of the whole enterprise:

If we could rise above the surface of the earth, and take in the whole country at a bird's-eye view, with visual power to discern all the details, the net-work of the telegraph would still be more curious to look upon. We should see a web spun of two hundred thousand miles of wire spread over the face of the country like a cobweb on the grass, its threads connecting every important centre of population, festooning every great post-road, and marking as with a silver lining the black track of every railroad. We should observe men, like spiders, busily spinning out these lines in every direction, at the rate of five miles an hour for every working hour in the year ... The whole net-work of wires, and the submarine cables which connect it with other equally active systems on the other side of the globe, are all quivering from end to end with signals of human intelligence.[14]

What is also interesting to note in this context is the misty-eyed, near-utopian tenor, also reflected in Gamble's reminiscence of the first intercontinental connection, that characterizes so much rhetoric around the concept of networks, both ancient and modern. The positive implications of the world being connected has long captivated the human imagination, and so often the implications are projected to be wholly positive. James Carey's famous article about the telegraph coined the term *electrical sublime*, a kind of infectious, near-religious enthusiasm that pervades thinking about wired networks.[15] But in that same essay Carey also notes a darker side: the connection with a certain kind of capitalist ideology, for the telegraph would increasingly reshape values according to the logic of the market, and "the telegraph via the grid of time coordinated the industrial nation."

Speed in communications and increased connectivity with other humans have the virtues of promoting efficiency, and sometimes they promote a greater understanding of others too. But other human values may be lost. As Carey points out, a uniform price system and a standard set of time zones were swept in with the telegraph (and the railroad). Local time and local pricing were gone. Nationalization—indeed, globalization—and standardization of formerly distinct regional cultures across the United States had begun.

Noteworthy, too, in the context of journalism, is the way in which the telegraph changed how citizens related to and even understood the nature of news. This new networked technology fueled a demand and a cultural expectation for near-instantaneous reports of events. More than a century and half later, we are in many ways seeing the logical conclusion of the telegraph with the mobile, Internet-connected phone revolution. People began standing outside of telegraph stations to hear the results of elections, battles, and other dramatic unfolding events, much in the way that people now constantly check their Twitter and Facebook feeds, email accounts, and news alerts on their smartphones.[16] The psychology of individuals across society and their relation to news was undergoing a substantial shift. This new mass demand signal also would substantially change what would be carried by newspapers themselves, which previously printed very little that was actually timely (in our contemporary sense of time). It also changed the style of journalism, putting an emphasis on compression of thought and brevity, a change that was "embedded in the architecture of the technology," as journalism historian Christopher Daly notes.[17]

The structure of the telegraph network, the types of information flows the technology favored, and the way that that information reshaped society were entirely governed by strong, often business-centric choices. The network reflected a hierarchy still. It was a structure of power and, so to speak, exclusion, even if nominally connected by "electrical bonds to the great national family," as telegraph engineer Gamble put it. This is a pattern that must be kept squarely in mind even when networks are architected for much greater inclusion. Indeed, it is precisely when they are designed for maximum inclusion, bringing with them the problem of overwhelming complexity, that solutions such as algorithms must be deployed.

The story of the telegraph, a tool now consigned to the trash heap of dated technologies, still serves as an important lesson in how all networks of networks, no matter how open in spirit and universal in aspiration, have an underlying structure that is fundamentally engineered by human choice. The telegraph is an archetypal reminder of this fact. The telephone system eventually began to replace the telegraph system as society's primary way of transmitting information quickly and efficiently from point to point, yet there is continuity between the two, insofar as they used analog, not digital, means to convey information.

Facebook: Engineering Six Degrees

The revolutionary moment in communications that is relevant to this book is, of course, the rise of the Internet. This brings us forward in time to the late 1990s and early 2000s, to the launches of Google and Facebook. Just as the intercontinental connection of the telegraph required a feat of engineering (as did the technology's initial invention by Samuel Morse and Alfred Vail)—making possible new commercial and cultural opportunities and expectations—these two behemoth Silicon Valley information companies would make their own consequential choices about network architectures. This has had substantial implications for, among other things, the practice and indeed the very identity of journalism.

The origin stories of Facebook are now shrouded in cultural myth and have been baroquely documented in numerous books and articles, as well as by Hollywood. The personalities involved, particularly Mark Zuckerberg, need little introduction. But I believe it is worth looking a bit more closely at the more technical network-related innovations that Facebook made, for in many ways they are underappreciated. As with the telegraph, Facebook's early engineering directions and computational strategies opened new cultural possibilities while limiting other alternative futures. Design and engineering choices interacted unpredictably with human consciousness and community values, bringing about new realities that we are only beginning to understand. In making these network choices and innovations, Facebook also changed the expectations of society and, in some ways, individual psychology relating to the nature of social relations—and therefore the spread of information.

As is well known, Facebook, or *thefacebook* as it was called when Zuckerberg launched the application from his Harvard dorm room in 2004, was not the first social media company. Both MySpace and Friendster preceded it as technology platforms for social connection. Friendster in particular provided the case study in what *not* to do from a technology and user-support perspective, and it was very much on Zuckerberg's mind throughout the start-up phase. The curse of success for Friendster manifested itself in routine and catastrophic site crashes as the number of users expanded.[18] In 2005, Facebook would see its own user base double between June and December, from three million to six million; the firm would spend $4.4 million on servers and networking equipment just that year. That astonishing growth created problems, existential ones.

At the end of that first full year, Mark Zuckerberg dropped by a class at Harvard (the vaunted Computer Science 50 course) to give an overview of his progress with the project. Recorded now for posterity, the lecture features Zuckerberg giving a technical account of how his operation avoided the disastrous "exponential" load on the servers supporting the site that brought down Friendster.[19] It may all seem like magic given Facebook's remarkably steady site performance now, but one of the most difficult computations that a social networking site needs to perform is figuring out how you are connected to other people. Solving for the shortest path between two people becomes exponentially difficult as you go out to friends of friends. For example, assume a network in which everyone has one hundred friends. To look for connections in the immediate network of your one hundred friends, an algorithm must look for matches among each of their one hundred friends of to see if there are connections. That's ten thousand pairs. The next "degree of separation" outward is then the cube of one hundred, or one million. "Friendster had large problems with this," Zuckerberg noted in the 2005 lecture, "because they were trying to compute paths six degrees out, or like seven degrees out."[20]

Facebook decided to distribute and split up the database according to clusters of friends, at first according to the college to which early Facebook users belonged (the network was initially restricted to universities). This could stop computation problems from becoming exponential. As the site began seeing one hundred million pageviews a day, with each page making numerous queries that draw on databases to serve up pictures and other tailored information, additional hardware and software challenges developed. But simultaneously the public demand for this sort of various precise edge or tie generation (recommendations of new friends) accelerated. As the engineering improved and the application became both better and more reliable, society began to reinforce it in a kind of human–computer feedback loop.

Further, as the Facebook network became larger, it became more useful; more of each member's friends were joining each day, providing the possibility of reconstituting a person's entire network online in a single place. This pattern roughly follows the well-known idea embodied in Metcalfe's law, whereby the value of a given telecommunications network grows in nonlinear fashion and is proportional to the square of the number of users.[21] Academics had long thought about targeted discovery in the human social graph, but there had been too many challenges to ever conceive of realizing its possibilities at scale. Engineering prowess had produced a new paradigm for connectivity, a product of mutually reinforcing computational and sociological trends.

Time and again, Facebook has tackled the next frontier of unconquered social network engineering challenges, and many of those early computational tasks are now child's play compared to the site's advances in recent years. What the Facebook team has done in the process is not just to create a new technological application to help users keep in touch with friends and stay connected, but in fact to seize on, illuminate, and make actionable a set of network "wires" that were hitherto invisible in human societies. Facebook has, for the first time, in a sense *electrified* the intangible web of human emotions and information. And by doing so, it has begun to change the very nature of this intangible web, as we can begin to observe and reflect on our relations and intellectual-emotional interconnections themselves. Even before the telegraph, there was a vast web of human connections that, through several degrees of separation, theoretically connected everyone from the Pacific to the Atlantic coasts of the United States. Yet these pathways were seldom utilized or made visible; they were seemingly random, mostly invisible, and certainly not precisely searchable.

When Zuckerberg founded his world-changing killer app at Harvard, he was heir to a certain tradition. On that very ground in Cambridge, Massachusetts, another network pioneer had, decades earlier, first made some insights into the nature of this vast invisible web of human ties. His name was Stanley Milgram, and his experimental work as part of the now-defunct Department of Social Relations at Harvard gave us the first real image of what the human social network really looked like. In the 1960s, Milgram got a $600 grant to conduct an experiment whereby randomly chosen people in Kansas and Nebraska would be asked to forward messages on to someone in the Boston suburbs. The results would become well known for illustrating the small-world phenomenon. Milgram found that there were between 4.4 and 5.7 intermediaries on average between the original, randomly chosen recipient of a postcard and the intended endpoint recipient. This was rounded to six degrees.

The six degrees idea was popularized by the John Guare play (1990) and 1993 movie *Six Degrees of Separation*, starring a young Will Smith. Malcolm Gladwell's *The Tipping Point* (2000) revived interest in Milgram's postcards experiment; in 2003, Duncan Watts published his own widely read treatise on the science of networks, *Six Degrees*. The echoes of Milgram were, in other words, reverberating across Harvard Yard when Zuckerberg, in essence, began to make Milgram's one-off experiments a kind of permanent feature of our world.

On January 23, 2004, Mark Zuckerberg and Harry Lewis, a computer science professor and the dean of Harvard College, began exchanging emails under the subject line "Six Degrees to Harry Lewis," a direct nod to the small-world idea.[22] A one-time student in Lewis's theoretical computer science course, Zuck, as Zuckerberg was known, had conceived of a computer application that would allow students to see how close they were to the dean on a social "map" of sorts. Lewis would serve as the *zero-degree* node— the fundamental reference point for all others in the social map. It was, seemingly, just a fun use of theory to explore the Harvard social community in a quirky way.

Zuckerberg explained to Lewis that "users would type in their names and see how many hops it took to be connected" to the dean. As Lewis has recounted, he then emailed Zuckerberg: "Can I see it before I say yes? It's all public information, but there is somehow a point at which aggregation of public information feels like an invasion of privacy." Zuckerberg showed it to him. Lewis recalls giving it a casual review and approving it: "Thinking it was another Friendster, I shrugged." In words that continue to take on historical irony, the dean then emailed Zuckerberg: "Sure, what the hell. Seems harmless."

In the early 2000s, Watts, then at Columbia University, along with coauthors Peter Sheridan Dodds and Roby Muhamad, reconfirmed Milgram's six degrees thesis using email chains, estimating there were between five and seven steps between email users across a sample of thirteen countries.[23] A 2012 study of Facebook found that there were about four degrees of separation on that network platform.[24] And in 2016, Facebook's data scientists determined that there are about three and a half degrees of separation between the platform's users.[25]

What Facebook states it is optimizing for (i.e., trying to accomplish for users) and what it actually does are two slightly different things. This must be kept in mind if we are to understand, as with the telegraph, what is in the network and what is excluded. Ultimately, the company wants to maximize engagement and time with the application. Notably, the goal of its algorithmic interventions is not necessarily to inform or consciously improve democracy, nor is it to strengthen relationships (in the way we might in the offline world by distributing our time and attention to sustain and enrich ties). Crudely put, the News Feed algorithm serves up what Facebook's engineers believe will maximize engagement on the platform, of nearly any kind (hate speech and calls to violence are exceptions), to sell that time and attention to advertisers. Of course, many of its executives espouse genuinely high-minded ideals about a better, more connected world. The company does use panels of thousands of human evaluators to determine whether different kinds of content are deemed to be meaningful, helping to improve the algorithmic curation. But the business strategy prevails on balance, and left out of the network are some human and democratic values. This tension was widely known to experts, but it took the 2016 election and the controversies over fake news to make these algorithmic issues known to the public.

The issue of values wider than just "engagement" and "relevance"—amorphous concepts that Facebook often falls back on in the company's public pronouncements—brings us back to journalism. There are two chief areas of debate with regard to Facebook and journalism: the business implications (the "platforms versus publishers" debate) and the effect on news consumption patterns (the "filter bubble" issue.) The first issue, which Emily Bell has continued to analyze for the Tow Center for Digital Journalism at Columbia University, comes down to a dispute over whether or not Facebook ought to be doing more to help news outlets stay in business.[26]

Facebook benefits enormously from the content of news organizations; stories of all kinds circulate on the network and make the environment more engaging accordingly. The company now has an enormous share of the advertising revenue that newspapers once saw; targeted advertising is better executed on the Facebook network, and mass media have had their advertising revenue reduced by billions. However, it is also the case that Facebook helps extend the reach of journalism institutions, allowing stories to reach many more people than ever. This platforms versus publishers issue may never be resolved in any final sense, although legal action and continued outreach on behalf of Facebook to help news media continue to complicate the relationship.[27] Facebook has also begun to fund original news content in video form, opening the door for new monetization models.[28]

The filter bubble concern rests on an argument that Facebook's algorithms are serving up content that tends to confirm people's biases and beliefs. Often oversimplified, this concern is a massively complex sociological issue and question, with dimensions that are still not well understood. Because studying information flows on Facebook at scale requires the company's permission, there has not been a lot of research to date. In addition, it is very hard to measure the overall rate of change because it is nearly impossible to determine how individuals' overall encounter with news— which, even prior to social media, was heavily influenced by offline social networks, geography, and more—is or is not being skewed.

Even the biggest, most comprehensive research studies published in the top journals have limitations. A large-scale study published in *PNAS* in 2017 found that among a sample of 376 million users, most tended to engage with a small number of news outlets. Those users who showed higher levels of engagement with news tended to interact with a smaller number of news outlets, suggesting a filtering effect. "Despite the wide availability of content and heterogeneous narratives, there is major segregation and growing polarization in online news consumption," the study's authors conclude. "News undergoes the same popularity dynamics as popular videos of kittens or selfies."[29] However, how these patterns fit into wider news consumption trends is not assessed in the study.

In 2015, members of Facebook's data science team published their own take on this problem in the journal *Science*; they tried to bring some wider perspective to the issue while admitting a modest amount of ideological filtering or skewing. "Individual choices more than algorithms limit exposure to attitude-challenging content in the context of Facebook," the researchers assert. "Despite the differences in what individuals consume across ideological lines, our work suggests that individuals are exposed to more cross-cutting discourse in social media than they would be under the digital reality envisioned by some."[30] That study saw significant criticism by other leading researchers for the sample that was used, as well as the way it posed its research questions and fitted data to them.[31]

One aspect that remains interesting about the 2015 Facebook study published in *Science* is the degree to which it highlights the fact that Facebook's own algorithms are somewhat beyond the comprehension of their engineers. They actually have to study their own algorithms to figure out what is going on; the computer code and the inputs of users interact dynamically, with no certain outcome that might be predicted by the creators of the platform. "You might imagine that they could just go into the next building and look directly at the code," computational social scientist David Lazer notes in an article that accompanied the original study. "However, looking at the algorithms will not yield much insight, because the interplay of social algorithms and behaviors yields patterns that are fundamentally emergent."[32] The use of machine learning and automation in regulating information flows on the social platform—part of the rise of "social algorithms" across the digital world—means that Facebook's process for filtering and selection remains a black box, both for the public and even, at times, for the company itself.

That said, it is possible that the appearance of increased homophily—"birds of a feather flock together," as the saying goes—is mostly an optical phenomenon, with Facebook just reflecting megatrends in American society toward greater group polarization and increasing selective exposure effects. What Facebook may actually be doing is making people's ideological preferences and their shared communities in some ways more visible and pronounced: we are shocked by how biased human behavior is (and political preferences are), and the social platform just allows it to manifest. Or perhaps not only is that true, but Facebook's algorithms are also genuinely accelerating polarization. Although big data studies will undoubtedly continue to examine the problem, and likely continue to produce conflicting findings, the one thing that can be said with some certainty is that there is a strong public perception of a problem. This perception may actually affect reality; the architecture of the network is constructed through code, after all, and can be adjusted. The growing chorus of worry will influence Facebook's business decisions. In the wake of the 2016 election and the controversy over highly partisan "fake news" that circulated on the platform, the company has already announced measures to tackle aspects of the overall problem, with algorithmic tweaks and efforts to monitor and combat deliberate misinformation efforts for political purposes.[33] Scandals involving the uses and abuses of user data, such as the 2018 scandal involving Cambridge Analytica, will further cloud the company's reputation.

What Zuckerberg and his team have done through their engineering efforts is not just to map and make visible the social graph of nearly two billion people around the world, but also to change the graph itself, rewiring it and bending it toward closer interconnection. It is in this way that engineered network architectures can not only allow for more links to be observed, but also, by allowing for that form of observation, change the very nature of the network, like some social Heisenberg principle. The visibility of the network continually reshapes the network itself.

This has had large, somewhat underappreciated effects on the public's relationship to news. There is a growing body of evidence that news shared through social networks, even online, can have pronounced and far-reaching contagion effects, influencing behavior, emotions, and offline actions such as voting.[34] Increasingly, news is not being consumed passively through a television set or newspaper, but rather with the important aid of friends and family, whose choices and opinions—a shaping architecture of its own kind—can make information more salient and emotionally impactful. Finally, Facebook facilitates the sourcing of stories in new ways, providing an instrument for people-finding, event detection, and community access (and indeed formation) like never before.

Google: Knowledge Graphs

As noted, networks have many forms and modes—from social to informational and mixes of both. No institution in history has done more in the area of information and knowledge networks than Google, whose well-known mission remains to "organize the world's information to make it universally accessible and useful."[35] That mission began a little over two decades ago, in 1995, with the meeting of two Stanford graduate students, Sergey Brin and Larry Page. Over the next couple of years, the two would imagine and then execute a new procedure for how knowledge should be valued in the world.

Search engines prior to Google had largely relied on the frequency of keywords on websites to serve up results to web users. AltaVista, once a popular search engine and now largely forgotten by history, operated in this way. This way of proceeding defied human needs and expectations for results in many cases. An example Page and Brin examined was typing the word *university* into AltaVista.[36] The first result was the Oregon Center for Optics, because the word *university* appeared multiple times in the headline of its home page. Such search engines were always susceptible to manipulation and gaming, and by the late 1990s they were being gamed increasingly by packing keywords into the source code of pages. Page and Brin also objected strenuously to the trend of advertisers paying for search results.

What became the world's most successful solution to the problem of search was developed in Page's and Brin's doctoral work, which initially was not even developed with the idea of creating a search engine.[37] Their ranking methodology system, or algorithm, was initially called BackRub, later to be renamed PageRank, after Larry. The idea, loosely modeled on the academic practice of citation, was to base authority and importance on the number of backlinks to a given website. This made use of the network structure produced by hyperlinking among web pages. "PageRank is an attempt to see how good an approximation to importance can be obtained just from the link structure," Page and Brin wrote in a 1998 paper they published, subtitled "Bringing Order to the Web."[38] In PageRank's most primitive, early version, a website would be evaluated by how many links it had, as well as how many links were directed toward that first set of directed links. This is done using a form of network analysis that solves for *centrality*, a mathematical method for determining which nodes, also called *vertices*, are most important or influential. Brin, a certified math genius from an early age, helped enormously to improve upon some existing formulas and techniques in order to apply graph theory and network analysis concepts to PageRank.

Google's ranking methodology is now supplemented by numerous other algorithms and factors from Google's massive search engineering team, including dozens of subtle qualities of websites and the velocity of information in real time on social media. It might be debated whether the idea of harnessing the wisdom of the crowd, whether through link analysis or supplementary indicators, is ideal in terms of identifying credible knowledge. Over the years, there have been numerous examples, some shocking, of Google's algorithms serving up bizarre or misleading results. Even in 2017, the company was continuing to deal with controversies of this nature—for example, ranking Holocaust denial websites very high for searches about the historic event.[39]

Yet it would be unfair on the whole to judge Google's failure or success based on its relatively infrequent missteps, which are often the result of an ongoing arms race with the deliberate manipulations of ill-intentioned actors. What is more consequential is what may be left out of search results at a structural level. The first page of results on any common query is inevitably filled with popular sites, which algorithms judge as being "useful" based on the frequency with which they are visited. Move over to Google Scholar (which filters only for research), however, and one can see what might be left out—namely, the vast library of the research world, which contains the most credible, peer-reviewed information. The irony here is that Page and Brin set out consciously to model the Google enterprise on the method of academic citation, but it is precisely the higher-grade but less-trafficked information that is frequently left out of the first page of search results. Google has improved this situation a bit with its Knowledge Graph project, which provides summary snippets of information at the top of search results (they are frequently drawn from Wikipedia entries). What Google has not yet done—and may never want to do, given that its chief goal is to maintain high volumes of usage from the crowd—is to privilege the world of peer-reviewed research and well-vetted empirical analysis.

Critics in recent years also have focused on the amount of data that Google is collecting about individual users while monitoring their browsing habits and targeting advertising and content to fit what the company perceives as the information most relevant to an individual user. The dangers here—first warned about by technologists Eli Pariser and Siva Vaidhyanathan, among others—are the siloing of individuals in echo chambers and contributing to political polarization, as well as the commoditizing of culture and knowledge. It is impossible to know, given that the algorithms used are all proprietary, how exactly Google is connecting nodes of information and knowledge together.[40] The signals it is using are complicated and subtle. Research on the degree of personalization in search results has shown somewhat modest differentiation among users.[41] What is indisputably true is that the world of online information Google is producing constitutes a dynamic system, with feedback loops that reinforce the deeper prevailing architectural choices made by the company.

The decision to delegate authority and order of importance to the behavior of the crowd has implications not only for what knowledge we find, but ultimately how we come to understand credibility and, to an extent, what we believe constitutes knowledge. The architectural choices of Page and Brin seem "democratic" insofar as they put the masses, to some degree, in charge of what's important. Yet we need not belabor the point, made time and again by theorists of democracy, that the majority can bring tyranny and that certain truths or rights must be protected by structures, legal or otherwise. Across many domains, from climate change to facts about how public policies are performing, the volatility of public opinion suggests a need for anchoring in the best science. No system of knowledge is perfect; the question is to what degree a system allows for imperfection. A computational solution certainly can improve upon fallible human processes and make retrieval exponentially more efficient. It is entirely possible that the rise of artificial intelligence will bring improvements in this regard. Google, to its credit, does use a system of paid human raters who evaluate the quality of search results.[42]

Google's aggregate effects on the news business are sprawling and largely hidden by being deeply embedded in a kind of structural shift in the way the public thinks about information. As with Facebook, there is a "platforms versus publishers" dynamic, whereby news organizations for many years have asserted that Google's aggregation of excerpts from news stories—in, for example, the Google News application—is a form of theft. Such claims have been fended off almost entirely in the United States (Europe is more complicated), and Google allows sites to opt out of being crawled and indexed. Google has consistently pointed out its upside for news institutions: it drives a tremendous amount of additional traffic to news sites, and it is the news organization's responsibility to monetize these larger audiences, not Google's.

There remain underlying tensions here. It is little remarked upon, but the overall utility of the web hinges in some part on the ongoing production of quality news. On almost any subject, search queries return useful and credible results because many stories generated by media organizations are available freely, or in excerpt form, online. Further, the underlying structure of Wikipedia rests in large part on the ability to cite and link to news articles. Look at the links in almost any substantial Wikipedia entry.

In this way, Google's Knowledge Graph, which keys off Wikipedia in many cases, rests on news content.

Of course, the news industry bears some large responsibility in deciding to put most of its content online for free in the first place, a decision that some consider a kind of "original sin."[43] As more content, such as video and music, is accessed through subscription services such as Netflix and Spotify, that original decision by news organizations looks less inevitable and wise. Many of the major newspapers have gone to metered paywall models accordingly, and they are actively experimenting with new models that lock down content for nonsubscribers. Still, given journalism's ambition to reach large audiences and inform the public, it is hard to see how creating many "walled gardens" of news content serves the wider values of either news media or democracy.

A phenomenon that is harder to capture is what the persistent availability of knowledge and information through search is doing to citizens' habits of news consumption—indeed, their underlying patterns of knowledge acquisition. This is a complicated area, and it has often been reduced to a fundamental debate over intellectual laziness, famously embodied in Nicholas Carr's 2008 article for the *Atlantic*, "Is Google Making Us Stupid?"[44] The appointment-driven "news habit" of consuming a daily newspaper and a nightly news broadcast has withered. We've gone from appointment-oriented to context-driven news consumption. If the patterns of the millennial and Gen Z generations persist, as discussed in chapter 1, those habits will be largely lost within the next twenty years or so. The availability of media in many forms, of many kinds, has broadened the media diet of Americans. Yet because people feel they can instantly access whatever they need to know, there is less felt need to be broadly informed on issues each day. The need to stay informed has changed as the *way* to be informed has changed.

We might lament this shift, but it would be strange if Google didn't affect how people approach and think about news. Leaving aside debates about alleged negative cognitive changes (i.e., Carr's) wrought by Google, it is perfectly rational for people to operate in this way. Recall political philosopher Russell Hardin's theory of the economics of ordinary knowledge: knowledge acquisition always involves a trade-off in terms of time, energy, and other opportunities lost. Google changes the public's calculus. Why bother to spend time acquiring something that can be accessed if needed? Undoubtedly, high-quality, targeted search changes this trade-off, and the public's relationship with news—which was once embodied in a commitment to daily consumption of a steady stream of a broad menu of news—is changing accordingly. The ability to draw on the world's vast library of knowledge instantly means it is less rational, using Hardin's idea, to access news content in a deep and even pattern. It is a golden age to be a news consumer, given the broad global menu available to almost anyone. With this, the habit of needing to consume news regularly in structured, appointment-driven ways has less force behind it.

A final word might be said about the positive changes in journalism practice that have been facilitated by Google. It is hard to find a single journalist who would claim that reporting is not better by having search engines. The ability to access background information and to retrieve source- and subject-relevant information on deadline is extraordinary, as compared to newsroom practice in, say, the 1980s. Productivity is enhanced substantially. News reports can much more easily build upon other reports, creating the capacity for context through practices such as hyperlinking. The capacity to fact-check—and be fact-checked—is massively increased by the architecture of search. The capacity for networked fact patterns to be brought together and reconciled (or at least debated) is a giant advance. It is true that the web has also enabled the spread of misinformation. But solely looking at the value for reporting practice, Google's search product (if used smartly) is a net positive for the efficient acquisition, verification, and expansion of information that can fuel journalistic output. Relevant data and research can be folded into so many more stories, likely making professional journalism in fact more accurate and contextual than ever.[45]

6 Data, Artificial Intelligence, and the News Future

A hundred years from now, what will the news needs of society be like? It is always risky to try to bring out a crystal ball, but there are certain things we know will stay relatively constant. In particular, we can count on culture to change—and, with it, forms of news.

Much of what we considered to be "news" in the past has been transformed into mere information by various technologies. Think of what the web has done to town meeting notices, weather updates, scores of ball games, and stock price fluctuations, and consider functions such as movie or local business reviews and how crowdsourced sites (e.g., Yelp or Rotten Tomatoes) have partially supplanted expert/professional media reviews. What we consider news now will likely become part of the architecture of information that streams through the Internet. Yet such transformations do not mean that news will then end or be fully given over to automation. Rather, news will move up the value chain of meaning.

Further, I will argue in this chapter that journalists may have an even more important role to play in an age in which original data is ever-more valuable. Rather than being a *gatekeeper*, the journalist of the twenty-first century will in part be a *locksmith* for sensitive data. But to understand why the machines are unlikely to take over the newsroom, we'll need to think a little more about the direction of change in terms of data and technology.

In my office at Northeastern sits a framed copy of the original front page of the *New York Times* on November 9, 1864 ("Price Four Cents"). The headline is "VICTORY!"; below it is a bit of Northern press jubilance, "Glorious Result Yesterday." President Abraham Lincoln had been reelected. The content of the six wide columns on the yellowing broadsheet is a dizzying cascade of data tables and microscopic print revealing the election results: "Details of Returns," "The Vote of City." Election counts for every single

office, from president down to canal commissioner, are detailed numeri-
cally in well-ruled columns. Ward-level votes are broken out in minute
detail. It is an explosion of raw and aggregated data, with no visualization
or interpretation beyond the obviously partisan headline. It is, to the modern eye, almost unreadable, or at least very overwhelming. It violates all of
our sacred contemporary design and user interface principles of reducing
clutter and drawing the eye to key points of entry. And, again, it is the
front page.

Now consider November 8, 2016, Election Day, the day that Donald J.
Trump was ultimately voted into office. The home page of the *New York
Times* website featured a real-time, data-driven application that presented
a "live election forecast." Three simple dials were shown for key indicators: "Chances of Winning Presidency," "Popular Vote Margin," and "Electoral Votes."[1] With a topsy-turvy day of election returns, many millions of
people were constantly checking this data-driven model through the day
and evening (and early morning the next day) to see the latest estimates.
The news experiment was something of a disaster as a predictive model;
the estimates proved to be wildly off. Although there were many questions
later about the model's validity—the day prior to the vote, Hillary Clinton
had an 85 percent chance of winning—it was nevertheless an interesting
experiment in doing things new ways, one that captivated a large online
audience. And it is surely a harbinger of new forms of news to come.

Even in 2016, news organizations eventually presented election returns
in tables, as they did in 1864, somewhere on their website or inside their
print editions—although any contemporary numerical tables are certainly
accompanied by graphics, interpretative language, narrative context, and
visual cues. Yet no contemporary news outlet would have ever even considered using the 1864 *New York Times* strategy up front, detailing data
tables across every column of a front page or home page. As a matter of
information design, the 1864 front page just slightly predates the *inverted
pyramid*, a newswriting structure for placing news-related information in
descending order of priority, subsequently well known to journalists of the
twentieth century.

As David T. Z. Mindich has argued, it was not until 1865 and the assas-
sination of Lincoln that the inverted pyramid began to emerge.[2] Looking
across the 152-year chasm between 1864 and 2016, we see two different

societies with distinctive news needs—news fulfilling a demand from society that met citizens' desires and understanding of what news should be, at that place and time. In decades and centuries past, much more purely "informational" content filled news media, from the shipping news to reams of stock price tables. News now has moved toward more "value-added," epistemological (building new knowledge about the world), or interpretative types of news stories and products. We expect journalists not only to provide context and relevant related information, but also to present it all in a format friendly to quick web browsing and busy, multitasking-filled lives.

As discussed in chapter 1, news media have a shape-shifting quality. "News is a historical phenomenon, always changing over time," Rasmus Kleis Nielsen has noted, "just as it is a socially contextual phenomenon, varying across space."[3] Yet there are underlying realities that connect even 1864 and 2016, as I will explain, that we must understand if we are to have any chance of thinking about the future of news. On Election Day 2168, what will the news needs of society be like? One fairly good bet is that it will involve machines and computation. The ability to process data and, increasingly, create meaning from inputs of various kinds is accelerating, leading to talk of potential breakthroughs in artificial intelligence in the coming decades. For years now, many things have been said about the potential for computers and artificial intelligence to automate and supplant work tasks, and even erase jobs *en masse*, across the labor market in the future. There are obviously great worries based on some of the projections.[4] Thousands of stories in the domains of financial and sports reporting are now being written by algorithms. There are predictions that perhaps 90 percent of all stories will be written by algorithms within a decade.[5] Where will this all go? Will it be a world of robot reporters? Beneath the classic fedora hat, will there only be silicon chips and wires?

Although the automation of beat writing in areas that are highly routine and quantitative—stock fluctuations, monthly earnings reports, soccer game scores, weather updates—is very real, the true impact of artificial intelligence on journalism remains largely hypothetical. The best thing we can do is to think as precisely as possible about what news is in a larger sense, then extrapolate how it might best meet up with the general trajectory of technology.

Artificial Intelligence in the Contemporary Newsroom

Experts make distinctions between various forms of automation and advanced computation and the field that is broadly called *artificial intelligence* (AI). The very definition of AI is something of a battleground even within the field itself. Further, the field of machine learning, technically a subfield of AI, has become so extensive and ubiquitous that the two are sometimes referred to synonymously. The idea of producing a general-purpose, human-like robot that could display flexible and convincing intelligence as we commonly understand it is a dream that persists across industry and academia. Yet general-purpose computer intelligence, what is called *strong AI*, remains a dream; the timeline of such a moonshot project could be very short or very long. For now, it is worth focusing on the more specific field of machine learning, which is being broadly applied across science, politics, business, medicine—and even certain dimensions of journalism.

As computer scientist Pedro Domingos notes, machine learning "takes many different forms and goes by many different names: pattern recognition, statistical modeling, data mining, knowledge discovery, predictive analytics, data science, adaptive systems, self-organizing systems, and more."[6] What, practically speaking, does this mean for the production of news? One obvious application is automated writing through tools and companies such as Automated Insights (a partner of the Associated Press, among others) and Narrative Science, which can quickly generate stories from structured data. Such automated writing is already happening. Other early applications in journalism include interactive chat bots, tools for optimization of headlines, selection of the best stories for social media performance, and data-mining techniques on large troves of documents. Yet these experiments are, to date, all somewhat limited. "Virtually the entire pipeline of news production is being touched in some way, to varying degrees of quality, by automation," Nicholas Diakopoulos, a computer scientist who studies journalism, notes. "These systems of course are brittle. ... They often only work in closely scripted scenarios and for very routine coverage and decision-making."[7]

Some of these applications and techniques will allow journalism to better meet audience demands in networks; it will allow the customization of journalistic products. Other applications, such as automated writing, will feel novel but will ultimately become mundane. As I have suggested, when technologies automate the collection and presentation of information, the cultural understanding of "news" will change. But these technologies also have the power to produce new forms of news in combination with well-trained journalists. "Computational journalism techniques such as multi-language indexing, automated reporting, entity extraction, algorithmic visualization, multidimensional analysis of data sets, [and] flexible data scraping," Amy Webb notes, "are allowing journalists to combine what they find in the data and then see the connections between facts, keywords and concepts."[8] The ability for computers to recognize images, sometimes called *machine vision*, and to be able to recognize and respond to human speech (e.g., Apple's Siri, Amazon's Alexa) through natural language processing (NLP) may also find important applications in both the reporting and audience engagement processes.

Given the challenges and possibilities inherent in our age of big data, there are likely two substantive areas where journalism augmented by machine learning can make an important contribution: (1) the monitoring of large social networks for relevant (and verifiable) information and trends; and (2) the analysis of large bodies of documents and data for hidden relationships. Many news companies are developing their own social media monitoring systems tailored to the needs of their reporters, or they are leveraging third-party applications from companies such as NewsWhip, which uses algorithms to track trending topics and monitor the flow of online attention toward certain stories.

Some of these algorithmic efforts to make sense of the giant stream of continuous social data are quite sophisticated, using data science to try to infer, from a statistical perspective, how combinations of subtle signals may indicate emerging and important news. Reuters, for example, has developed a tool called News Tracer, which assigns a "newsworthiness score" to nascent events developing on Twitter, potentially allowing reporters and editors to get to stories earlier in their life cycle. "With the proliferation of smartphones and social media, it means that there are [a] lot more witnesses to a lot more events," Reginald Chua, the executive editor for editorial operations in the data and innovation division of Reuters, says. "We can't be at everything. Our tool helps shift some of the burden of witnessing and lets journalists do much more of the high value-added work."[9] In this way, the use of cognitive technologies does not so much replace the journalist as free him or her to work on contextualization and interpretation of emerging information.

News Tracer is a powerful example of how humans and computers together—*augmented journalism,* as some are calling it—can combine to make insights that neither could alone.[10] The algorithm works off a knowledge base of credible sources, a dataset that is initially selected by Reuters journalists. This seed dataset is then expanded by algorithms that can trace out larger, interconnected networks that are also likely reliable. The idea is that credible individuals tend to follow other credible sources and institutions. The News Tracer software then performs the analytical task of looking at how a given piece of information, or tweet, is cascading through the feeds of various people on the network (e.g., is it being rebroadcast by other reliable accounts), as well as the identity and location of the account that originated the information.[11] To improve the News Tracer model, the data science team at Reuters has continued to perform computational social science research on issues such as prediction of the news value of natural and manmade disasters, as well as the debunking of rumors on Twitter.[12] The news company has claimed that News Tracer has beaten global news organizations to important stories dozens of times, giving its reporters a six- to eight-minute advantage.[13] However, algorithms so far have not proved as effective at debunking false information, maintaining the need for humans to stay in the loop.

Further, as mentioned, machine learning can be used to help superempower journalists performing investigations that involve large troves of documents and datasets, allowing for work to be completed that would otherwise take much more time and staff resources.[14] The *Atlanta Journal-Constitution*'s 2016 Doctors and Sex Abuse series began with a reporter noticing a strange pattern of physicians being allowed to practice and keep their licenses despite being sanctioned for sexually violating patients. When the newspaper's team wanted to investigate whether this was a national trend, they wrote *web scrapers,* or automated programs, that collected data from state medical board websites. This yielded more than one hundred thousand documents. They then used machine learning to vastly reduce the number of documents that should be directly reviewed by reporters, letting the algorithm look for certain keywords, which then gave each document a "probability rating that it was related to a case of physician sexual misconduct."[15] With their final dataset reduced to six thousand suspicious cases deserving of direct human inquiry, the *Atlanta Journal-Constitution* was able to pull off a path-breaking series and punch above its weight as a regional news organization without a large data journalism team.[16]

Another important machine-learning-involved use case in journalism is the testing of the veracity of sources, such as seeing if government officials are putting out valid or skewed statistics. For example, the *Los Angeles Times* used these computational techniques to demonstrate that the Los Angeles Police Department was misclassifying minor assaults and downgrading serious assaults to make its crime-reduction efforts look better. The *Los Angeles Times* team used machine learning and built a classifier—a model that allows algorithms to sort cases into different "buckets" depending on their relevance—to evaluate data from 2005 to 2013, obtained through public records requests. The team described their computational method thus: "To conduct the new analysis, [the] *Times* used a machine-learning algorithm. The computer program pulled crime data from the previous Times review to learn key words that identified an assault as serious or minor. The algorithm then analyzed nearly eight years of data in search of classification errors. Reporters refined the algorithms and selected a random sample of nearly 2,400 minor crimes from 2005 to 2012 to determine their accuracy. The sample was stratified by crime categories and the margin of error was plus or minus 2%."[17] The reporters were able to analyze more than four hundred thousand incident reports—a task ordinarily impossible given the realities of media deadlines and limited available time in newsrooms. Employing a sophisticated strategy that resembles academic research—and at the same time reduces the need for tedious, rote document review work by humans— the team ultimately used these computational techniques to perform an important act of watchdog journalism, of public accountability, with regard to a powerful government agency.

It is possible that computers will become wildly creative in the future, but it will be difficult for any algorithm to bridge what is technically possible and what is desired by public audiences in the fluid symbolic context of human culture. The creativity involved in generating original story ideas is a function of a moving variable: audience needs and societal context. For such back-and-forth processing, the human brain will remain uniquely positioned. "There's no such thing as a machine that comes up with story ideas, and spits them out," Meredith Broussard, a journalist and researcher who studies artificial intelligence, notes. "You don't actually want that, either, because computers are insufficiently creative. A human investigative journalist can look at the facts, identify what's wrong with a situation, uncover the truth, and write a story that places the facts in context. A computer can't."[18]

Journalists as Society's Locksmiths

It is well worth noting that in the pioneering investigations by the *Atlanta Journal-Constitution* and the *Los Angeles Times*, the chief operational problem being remedied was a lack of government transparency. In the case of the doctors who had committed sexual abuse, the data were theoretically available on state medical board websites but were not interpretable for the public, with patterns being hidden in the mass of bureaucratic forms; in the case of the Los Angeles crime data, the methods used for data labeling by police were obscure, and the records themselves were not easily available for inspection. This is a kind of "market failure," in which society itself does not provision the required information and the necessary knowledge-generating institutions. Journalists often must make up for the deficiencies of democracy. Even if the data in question were more open, there remain problems of data integrity and public comprehension. As the saying goes, data doesn't interpret itself. Moreover, any dataset (the kind that is readily released by institutions) is more likely to put a positive spin on the activities of its creators, obscuring truth and genuine knowledge, as was the case in Los Angeles.

No matter how good cognitive technologies get, they will require specific kinds of valuable (and often sensitive) data to make useful insights in the public interest. "The potential for AI to augment the work of the human data journalist holds great promise, but open access to data remains a challenge," notes a report by the Tow Center for Digital Journalism at Columbia University. Even as the number of datasets openly available on the web increases exponentially, many kinds of data will remain locked down. In fact, there might be every expectation that powerful institutions, governments and corporations especially, will continue to make it difficult to obtain and interpret all manner of data that might provide accountability.

There is no better example of this than the dynamics that unfolded during the administration of President Barack Obama. The White House developed a number of substantial open data initiatives; at the same time, journalists and watchdog groups consistently complained that the material they really wanted—requested through Freedom of Information Act requests and regular contact with government staff—only became more difficult to obtain over that period.[19] A 2017 Knight Foundation report that surveyed more than two hundred experts found that four in ten said access to federal records had become worse between 2012 and 2016, and nine in ten experts expected the problem to accelerate in the coming years.[20] Indeed, observers saw the development of further problems and slowdowns in responding to requests during the Trump Administration's early stages.[21]

The idea of locksmithing as journalism's primary role is in keeping with an idea articulated by Daniel Kreiss, who has argued that as

an ideal, we should normatively value journalism as a form of institutionally organized "civic skepticism," where journalists exercise scrutiny over elites and institutions, seeking to hold them to account for the democratic values of the civil sphere—equality, liberty, and justice—through their literal and symbolic control over the publicity of the powerful. While journalism often fails to live up to this ideal, valuing civic skepticism necessarily recasts the debate over journalism's future—from an emphasis on correcting for market failures in the provision of information, to a focus on the value of a strong and enduring institution that expressly serves the democratic function of holding power to account for the values of the civil sphere.[22]

Careful and sustained civic skepticism also has the virtue of being an area in which there is less direct market competition online. It is an approach to information that requires creativity and novel connections among disparate informational nodes. Civic skepticism is not focused on the commodity public information and data that the likes of Google can simply vacuum up and repackage. Well-supported, evidence-based investigative and accountability work takes time and patience that typically only individuals and organizations dedicated to such work have.

There is nothing natural or inevitable about where sensitive data ends up—hidden away or spilled into public view—even if the web theoretically facilitates more leaks. Indeed, most of the major leaks—Wikileaks' various disclosures, the NSA revelations by Edward Snowden, the Pulitzer Prize–winning Panama Papers series that unearthed offshore banking malfeasance[23]—have all been shepherded into the public eye by news organizations. Without news organizations to vet the information ethically and critically, the leaks process loses its social utility and becomes susceptible to gaming and disinformation campaigns. Further, leaks themselves might become less common without news organizations because the primary incentive for leaks is publicity. This is all to say that even in an era of automation and big data, journalists retain their importance. Further, as open data becomes freely and universally available, the data that cannot be readily obtained increases in its value (both social and economic).

Obviously, NGOs, watchdog and advocacy groups, activist citizens, and researchers of many kinds will also continue to do important public-facing work that complements journalistic work in this area. But journalists are typically the ones who are holding the data- and document-bearing institutions accountable *as systems*. Journalists and their work have been involved in, and indeed have driven, many of the major public policy shifts toward greater openness in government and society more generally.[24] Increasingly, many argue, what is required is not more data for its own sake but "targeted transparency" that facilitates sense-making and allows for citizens to evaluate the performance of government on important matters.[25] Journalists are society's great locksmiths, without which machine-learning algorithms are only of limited use. Only together can computers and humans truly perform watchdog functions that will have enduring social value.

In an age of greater complexity and more open data, one of the great challenges for society will be ensuring that what looks like transparency is in fact real transparency, and not just its appearance. Websites, open datasets, visualizations, and statements of openness can make institutions look as if they are transparent when in fact they are not. Journalists must play a crucial role in not only getting one-time disclosures and datasets but forcing systems to be more systematically open.

State Financial Disclosure: A Case Study

There are many domains of government in which, even now, data is hardly open or free and is not susceptible to algorithmically derived insights. The following is an account of one such domain—an area that is especially important for the functioning of democracy. It is an area in which networks of recognition likely only can be formed through very tough locksmithing by journalists, who will need to request, analyze, and interpret sufficiently to generate public discussion.

State-level financial disclosure systems now exist across the United States. These are the rules and regulations that require politicians, officials, and candidates for office to disclose their personal finances to guard against conflicts of interest and corruption. Broadly speaking, Americans have steadily professionalized and institutionalized ethics measures in the public sector over the past half century. This is part of a broad shift that swept through many areas of society beginning in the 1950s and 1960s and which has been characterized as the "rise of the right to know."[26] Ethics commissions and officeholder and candidate disclosure laws were borne out of the post-Watergate reform era in the 1970s; a second wave of states set up commissions in the 1990s.[27] Although almost no states require the release of tax returns, as is frequently customary (but not mandatory) in federal campaigns, many have put rules in place requiring candidates and officeholders to disclose their income, assets, and other ties germane to the integrity of office.

Along with my Northeastern colleagues Mike Beaudet and Pedro Cruz, I embarked on a quest in 2016–2017 to figure out how exactly the varying state policies for financial disclosure compared with one another.[28] The decline of newsroom headcounts and the corresponding decimation of reporting ranks at the state and local levels likely mean that there is far less ability to turn data into knowledge, even if government data, much of it trivial, increasingly becomes available as open or big data. It is worth bearing in mind the 2009 prediction of David Simon, creator of *The Wire* and a former *Baltimore Sun* journalist, that the "next 10 to 15 years will be halcyon days for local corruption."[29]

How truly "open" are many of these ostensibly open state-level policies? We set out to look at what might be considered the core of ethics-related transparency data at the local level: the information listed on the personal financial disclosure form and the practices and rules associated with it. These forms are filed annually with state ethics commissions or agencies by public officials such as legislators in almost all states (Idaho, Michigan, and Vermont are notable exceptions), as well as by candidates for public office. The public hears a lot about campaign finance disclosure, but less about these types of personal income and assets filings, and they are seldom accessed. In theory, these forms cut to the bone: they may not be tax returns, with all the intimate detail of IRS 1040 forms, but they should provide the public with a sound overview of how people with political power make and manage their money.

What we found in looking at these forms and analyzing state laws, however, was a set of widely varying practices, with relatively few states performing well on overall transparency measures when we reviewed the

forms side by side and scored them on fixed criteria. In fact, about 80 percent of states rated poorly.

New Hampshire's single-page form scratches the surface, but barely. Filers are simply required to check a series of boxes without disclosing any
actual numbers. Wyoming's form is only slightly more enlightening. Officials must indicate if income was earned through security or interest earnings or real estate, leases, or royalties, but otherwise are not required to
provide more detailed information. In Arizona, filers must only indicate
compensation, personal debts, and financial interest in trusts or investment
funds over $1,000, without disclosing the real figures. In Missouri, sources
of income over $1,000 for the filer and family members must be listed, but,
again, without actual amounts.

The lack of specifics makes it difficult for the public to evaluate perceived
or actual conflicts. This is a running theme throughout many of the disclosures, as there is no standard across the states requiring either specific
amounts or detailed monetary ranges. Complicating matters is the seemingly obscure disclosures some public officials are making on the forms.
In Nebraska, for example, Governor Pete Ricketts' less than revealing disclosure in 2015 did include a detailed twenty-five-page gift list. Those gifts
range from a framed picture of high school friends and a quart of ice cream
to a vase with two carnations and two large jars of gumballs. It is questionable whether disclosures like this serve the public good or do anything to
keep our officials more open and honest.

Another consistent finding was that there seems little capacity or authority in the state ethics commissions to verify information, even on a random
basis. Most states that do audit the forms are only checking to make sure
they have been filed on time and/or filled out completely. When actions are
taken against filers, it is usually for not filing a disclosure.

We also looked at the scope of disclosure—who has to disclose, and
how far down the bureaucracy rules extend—although the lines get blurry
pretty quickly in many places. In states such as Maine, Colorado, Nevada,
and Ohio, there are relatively few filers, with only major officials and/or
candidates required to disclose. But other states, such as Florida and California, see tens of thousands of annual filings, and even some states with
much smaller numbers of state employees, such as Massachusetts, Oregon,
Oklahoma, and Washington, require filings from more than four thousand
people. Based on our review of hundreds of political ethics-related news

articles and court cases, this area is worthy of further study because lower-level aides and officials are often caught up in ethics scandals.

Equally important is the ability for the public to access whatever information is disclosed. Once again, there is no standard. Nearly half the states received top scores in our evaluation by providing easy online access to information; such states included a diverse mix—for example, Alabama, West Virginia, Alaska, New York, and New Jersey. However, notably, not all of those states performed well overall in our transparency scoring, as the forms themselves revealed relatively little.

Other states, such as Maryland—where requesters must show up in person at the state ethics commission and present identification to obtain a copy of a disclosure—and our home state of Massachusetts—where an ID must also be provided, even to gain online access—place hurdles for members of the public to jump over to access what is supposed to be public information. Both Maryland and Massachusetts also notify filers that someone has requested to look at their disclosures, a policy that could create a chilling effect on open government and transparency efforts. About fifteen thousand public servants in Maryland file annual financial statements, but only a few dozen of those files are accessed each year by the public, according to the Maryland State Ethics Commission. In 2015, just fifty-eight requests for financial disclosure forms were made in Massachusetts.

The need to police the transparency of systems will, as mentioned, grow increasingly more vital as the complexity of the online ecosystem grows. Louis Brandeis famously wrote, "Sunlight is said to be the best of disinfectants." Forgotten is the subsequent clause, "electric light the most efficient policeman."[30] The reality is that even the most wired of corruption-policing regimes may be hampered when there is little data to illuminate. Disclosure is surely no panacea, but its absence may be less likely to produce confidence or ethical behavior in the system. Ultimately, our research findings highlighted the extent to which democracy needs to refocus on *targeted* or *core transparency*—measures that are vital to the integrity of the public system and the public interest.[31]

Meaningful data are not always yearning to be free. Long into the future, there will need to be someone (likely a journalist) to force the right data into view and nudge the machine-learning algorithm in the right direction. To facilitate networks of recognition centered on difficult questions and generate the social capital necessary for democracy, there will first need to be a lot of facts surfaced by the journalist to generate deliberation and conversation among citizens, governments, and NGOs. Such locksmithing is painstaking work, and it is of little consequence if it cannot be translated into fuel for user groups, stakeholders, and, ultimately, network formation. Technological optimists may boast of the ability of increasingly sophisticated machine-learning models to identify patterns and produce meaningful insights, but all such models require training data: without rich data across many domains, all the AI in the world is of limited help.

7 Journalism's New Approach to Knowledge

Journalism must become much smarter. In a world dominated by social media—a world seemingly inundated by frothy, viral content—this can seem almost counterintuitive. But a greater embrace of knowledge is essential if journalism is to retain its value, both socially and economically. To embrace this new knowledge-driven mentality, though, journalism needs to get past some enduring tensions between informing and engaging. These tensions are generally framed by what is called the *Lippmann-Dewey debate*.

Almost precisely one hundred years ago, Walter Lippmann, America's most influential early media critic and political journalist, penned a series of bleak, powerful essays that he assembled in a volume called *Liberty and the News*. For the previous five years, Lippmann noted, over the course of World War I and its aftermath, societies had been mobilized toward a central, all-consuming cause, one that had suppressed truth in the service of victory and "conscripted" public opinion. The wholesale distortion of truth, the retreat from a commitment to veracity and the headlong plunge into propaganda on all sides, had had damaging consequences. "The work of reporters has thus become confused with the work of preachers, revivalists, prophets and agitators," Lippmann, who as a young political and military aide had helped fuel the propaganda efforts, wrote.[1] And without a "steady supply of trustworthy and relevant news," the survival of government by consent was in grave doubt.[2]

That was the world as he saw it in 1920. Lippmann was not alone among intellectuals in noting that a new mode of operation and a reorientation toward an increasingly industrial world was, at that moment, becoming vitally necessary across many domains. The wrenching shock of that global war—the millions it killed between 1914 and 1917, the religious and moral

crisis it provoked, and the destruction of civilizations that had existed for
centuries—prompted artists and writers to invent new forms of expression,
embodied in the modernist movement that flourished in novels, poetry,
music, and painting. The creative arts had to find a new vocabulary to articulate the overwhelming confusion and complexity of this new, fragmented
reality. A new basis for understanding the world was required.

Likewise, in the realm of public affairs, Lippmann noted, questions
of policy, regulation, and governance had become so dizzying, so extensive and baffling, that they were no longer intelligible to average citizens.
"Everywhere to-day men are conscious," he wrote, "that somehow they
must deal with questions more intricate than any that church or school had
prepared them to understand." Thus, journalism must reorient itself and
build a new foundation for popular governance, for "in an exact sense the
present crisis of Western democracy is a crisis in journalism."[3] The world
was becoming complex, and journalism was in the hands of persons who
were not up to the task of covering the world. It was very much a "factual
recession," with parallels to our own moment of crisis.

Over the next decade, both Lippmann and Dewey, his famous counter-
part and America's most prominent philosopher at the time, would publish
profound books concerning the idea of a "public" in crisis: Lippmann's
Public Opinion (1922) and *The Phantom Public* (1925), Dewey's *The Public and
Its Problems* (1927). Through the years, Lippmann has been cast as the pro-
ponent of a top-down approach to informing of the public, whereas Dewey
has been extolled as the champion of more bottom-up modes of generating public knowledge. Their ideas remain touchstones concerning the
relationship between news and democracy, and in many ways we remain
in the grip of their ideas and frameworks. From that period of crisis were
born many of the central questions with which communication scholars
and journalists still grapple: To what extent is journalism responsible for
informing a citizenry? How much should the citizenry actively participate
in setting the media agenda or guiding attention?

Binary frameworks ultimately prove brittle and overly crude, especially
when reduced to clichéd abstractions such as "elites" versus "masses,"
experts versus the wisdom of crowds. Michael Schudson argues that, in fact,
the characterization of the Lippmann-Dewey "debate," which frequently
identifies the "elitist" view with Lippmann and his emphasis on the need
for expertise in a complex society, rests on a misguided reading of *Public*

Opinion (1922) and *The Phantom Public* (1925).[4] "The intellectual challenge is not to invent a democracy without experts, but to seek a way to harness experts to a legitimately democratic function," Schudson notes. "In fact, that is exactly what Walter Lippmann intended."[5]

From that so-called Lippmann-Dewey debate in the 1920s that catalyzed these competing notions all the way to the present, this framework has implicitly continued to color and influence media and communications discourse. It is our intellectual inheritance in journalism, and it remains the framework for contemporary debates in this space.[6] The rise of the blogosphere and social media platforms has added fuel and complexity to the debate. Is journalism about engaging citizens or informing them? Might citizens, now endowed with individual broadcast power, just perform this function themselves and jettison the intermediary?

For too long, the perceived conflict has likely been overstated. In his original review of *Public Opinion*, Dewey praises Lippmann's "brilliancy," although he suggests that "Mr. Lippmann seems to surrender the case for the press too readily—to assume too easily that what the press is it must continue to be."[7] Lippmann's chief target was the political party machines that were threatening to overwhelm public interest and reformist efforts of that era. It was not that Lippmann opposed citizen involvement in democracy.

It is a lesson worth remembering. There is always another zone of power that should be watched, by whatever means, with greater suspicion than the power of the ostensibly gatekeeping press: the organized groups both political and corporate that are battling for influence and attention. After our recent decades of fierce debate over how much professional journalism should be given over to crowdsourced, citizen-centric journalism, the 2016 election and its aftermath have awakened many observers to the fact that the elites versus masses debate in media needs updating. The dangers to democracy are deeper than media, although media are deeply implicated in both problems and solutions.

The implicit argument in this book is that we have reached a moment when the elite-populist dichotomy is no longer useful in helping us think about the future of journalism. Such binary terms obscure the key emerging opportunity for journalists in the digital era. We are seeing the extraordinary and simultaneous transformation of two important and interrelated spheres: the democratization of media power and the democratization of knowledge. It is now possible to conceive of a journalism that is actively engaged with citizens on digital platforms—one that even consciously facilitates the formation of online publics—*and* a journalism that sees its primary value as connecting citizens to wider pools of knowledge, whether research, public records, or systematic data. However, the ability to engage constructively, to choose between the myriad options, perspectives, and opportunities in the online infosphere, itself requires greater knowledge on the part of journalists. The sine qua non, then, of effective networked journalism is knowledge, without which journalism is destined to make the mistakes of the past.

The debate over journalism's future is increasingly not necessarily a matter of engaging (populist) versus informing (elitist). Journalism in the era of expansive digital connectivity has the potential to create a kind of "virtuous circle" involving journalism and democratic citizenship, fueled and framed by knowledge. Both Lippmann and Dewey saw knowledge as the key. Dewey said the future of democracy depended on the "spread of the scientific attitude," particularly as a bulwark against propaganda.[8] Lippmann also located the solution to democracy's ills in the creation of more social-science-oriented institutions.

The linchpin in this philosophical turn is a much stronger journalistic grasp of issues. Theorists of what has been called *deliberative democracy* stipulate that decision-making is legitimated when citizens give morally justifiable reasons for their views, fostering an iterative process of exchange.[9] In this context, James Ettema has noted that journalism "must itself be a reasoning institution that aggressively pursues, rigorously tests, and compellingly renders reasons that satisfy the key criterion of deliberative democracy."[10] Without this new commitment to the aggressive application of knowledge and reason in the public sphere, journalism risks replicating old patterns (e.g., "he said, she said" stories), even if dressed up in more populist and engaged form.

To organize action on issues of increasing complexity, citizens must have a basic level of understanding, rooted in substantive knowledge. Journalists can facilitate this understanding in targeted ways. As mentioned, this potentially creates a virtuous circle: engaged citizens are more likely to be informed because interest incentivizes learning. By understanding the true stakes in issues for average citizens, journalists are therefore more likely to engage the public. Such a philosophical turn toward a synthesis of informing and engaging is a vital step forward toward centering news practice on networks of recognition—and dealing squarely with the hybrid domain of facts and social facts, in which ordinary citizens increasingly live and try to make sense of the world.

The Crowd and the Library

Shifts toward a world of networked knowledge have implications in two directions for journalists. First, journalists must be more highly engaged with public audiences to discover more "horizontal" knowledge, as anecdotal perspectives and related information are now part of the broader picture that is necessary to "know" a subject. Second, in a world replete with misinformation and a vast sea of noise, journalists must have much deeper subject area expertise in order to engage with audiences and help build well-grounded public knowledge. Because selection now becomes so important, knowledge must be built into the journalist's training so that selection is judicious and intelligent.

Both the crowd and the library are part of the continuum of knowledge now, and if journalists are to succeed in this new world, they must learn to thrive in both spaces. The *crowd* is the world of social networks and social facts, of nonhierarchical perspectives and anecdotes. The *library* no longer looks like it once did; it's hypertextual, after all, linking to near-infinite sources. The growth of Wikipedia represents this new hybrid paradigm. Wikipedia harnesses the crowd but still uses systematic information as some of its core building blocks, an idea more congenial to that of the traditional library, or systems of scientific and research-based knowledge.

Ultimately, deep investigative and explanatory reporting can facilitate important kinds of network formation that would not exist otherwise. Networks of recognition may require very deep digging to come into being. In previous chapters, we examined instances in which journalists engage with crowds online. Let's now look at several examples of stories that can help explain how journalists are engaging with and creating more systematic knowledge too. Such stories provide powerful fuel for networks of recognition. They show how different levels of society are implicated in common concerns and how they are knitted in core citizenly responsibilities and decisions.

"Hell and High Water"

Journalists traditionally have been "translators" of knowledge, writing through medical studies and government reports, making arcane scientific and social-scientific findings more accessible for the lay public. Yet new roles, even higher up the information value chain, are emerging across the spectrum of journalism, powered in particular by some digital-native outlets and nonprofit news organizations. This trend is in part fueled by accessible software that facilitates computational and visualization tasks, but it is also spurred by an emerging spirit of daring and confidence—venturing through areas into which traditionally only academics and government agencies could or would go.

When Hurricane Harvey blasted the port city of Houston, Texas, in late August 2017, the flooding and mayhem unleashed seemed, to many residents, policymakers, and news reporters, a shocking event. Many dozens of lives were lost, and more than $100 billion of property and infrastructure damage was wrought. Yet the events should not have been shocking, necessarily.

Why? The *Texas Tribune*, ProPublica, Reveal, the University of Texas at Austin, Rice University, Texas A&M Galveston, and Jackson State University had combined forces to produce a stunningly prescient, interactive story roughly eighteen months prior called "Hell and High Water," which presented the likely scenarios for a direct hit on the city by a major hurricane. Calling Houston a "sitting duck," the story—by journalists Neena Satija for the *Texas Tribune* and Reveal, Kiah Collier for the *Texas Tribune*, and Al Shaw and Jeff Larson of ProPublica—showed how citizens and policymakers had utterly failed to prepare for the kind of storm that, almost inevitably, would eventually come to the city.

Using accessible visualizations based on sophisticated modeling by their academic researcher partners, the journalists constructed a compelling warning to the city of Houston. Academics spent many hours helping the journalists bring together the data files and render them accurately. The legacy of the story—the great unheeded warning it represents—is in many ways tragic, but it also stands as an important case study in how journalism can team up with research institutions to produce powerful public-interest reporting.[11] The journalists served as both knowledge- and policy-brokers, or translators, mediating vital technical information for the public. But they did more than just translate. The mix of policy reporting and complex data-visualization tasks combined to create an altogether new body of knowledge, one informed by expertise but accessible enough to average persons. It has continued to contribute to a substantial public conversation about coastal development, zoning, planning, and climate adaptation. In a century when risk and resilience are becoming part of our vocabulary—from the global climate to financial markets to networked technologies, danger seems to know no boundaries—such reporting based on forecasting will likely become essential in terms of preparing the public to make smarter adaptive choices that can lead to resilience.

"Fatal Force"

The answers to society's public policy problems are not always sitting on the servers of government or academic institutions, and sometimes the data needed to effect positive change is simply not available. There are times, increasingly, when journalism must serve in the role of primary knowledge generator, especially when governments and markets fail to provision adequate information on vital policy issues. Such was the case with police-involved shootings in the United States—which, extending through many other high-profile cases, such as the death of Michael Brown in Ferguson, Missouri, became the focus of roiling national debate.

Motivated by the fact that the government was not collecting systematic information on fatal shootings by police officers, the *Guardian* ("The Counted") and the *Washington Post* ("Fatal Force") both stepped in and decided to function themselves, in essence, like administrative government data-collection agencies.[12] Both used methods of crowdsourcing and targeted records requests, as well as broad interviewing, to compile unique databases. The *Post*'s accounting of fatal shootings by police officers nationwide, in particular, embarrassed the FBI and spurred action. It prompted FBI Director James B. Comey in October to call for better federal recordkeeping and greater police accountability. "It is unacceptable that the *Washington Post* and the *Guardian* newspaper from the United Kingdom are becoming the lead source of information about violent encounters between police and civilians. That is not good for anybody," then-director Comey told law enforcement officials and politicians.[13] The agency's new method and approach to data collection in some ways replicates the *Post*'s.

The news organization–generated data was referenced across Capitol Hill in testimony, and it was cited in various academic and nonprofit reports, adding to the overall body of knowledge about this important public policy issue. The data collection itself has helped to grow and sustain networks of recognition, uniting grassroots activists and policymakers in a vital discussion about mutual obligations to all citizens and issues of trust among communities.

"Crime in Context"

This data-collection role, sometimes discussed as the building of news apps or online interactive databases, can put news organizations themselves in the position of knowledge generator. The sophisticated interpretation of the data is another matter, too, and other news organizations such as the Marshall Project, along with a suite of other digital-native outlets such as Vox and FiveThirtyEight, have risen to that challenge.

To take just one example, journalists Gabriel Dance and Tom Meagher of the Marshall Project set out to answer the question of whether violent crime was up or down in the United States, a political football of a question that seemed highly susceptible to bias and distortion. In a penetrating feature, "Crime in Context," the journalists carefully collected and synthesized four decades of FBI data from sixty-eight urban police jurisdictions, assembling a nuanced portrait of the conflicting macro- and microtrends playing out across the United States in cities.[14] Using statistical techniques to create weighted averages and smooth random fluctuations, the piece featured data visualizations to explore the complex nuances of this ostensibly simple question—"Is violent crime up or down?"—getting beyond the political rhetoric that so often distorts public perceptions of safety. "We found that the reported violent crimes rose in our cities last year to its highest point since 2012," the journalists conclude. "But viewed in the broader context of the past five decades, crime remains near record lows."

"Machine Bias"

In the age of big data and increasingly accessible computational tools, there is also the opportunity for journalists to become hybrid journalist-academic researchers themselves, building upon academic literature to study thorny public-interest issues. Such was the case with ProPublica's "Machine Bias" investigative feature, which explored and critiqued computer software that scored criminal defendants' likelihood to reoffend. The software was being used by courts to help shape sentencing decisions.[15] That pioneering investigation of an algorithm, what Nicholas Diakopolous has called the emerging journalistic beat of "algorithmic accountability," stands as a tour de force of fearless and sophisticated reporting in the public interest.[16] Building upon the existing academic literature that had parsed and critiqued other risk and sentencing software, reporters and editors Julia Angwin, Jeff Larson, Surya Mattu, and Lauren Kirchner performed their own statistical analysis, concluding that the software being used was biased against minority defendants.[17] Baked into these ostensibly neutral algorithms, developed by a private company and deployed by the justice system, were the unmistakable signs of discrimination, going both ways: "The formula was particularly likely to falsely flag black defendants as future criminals, wrongly labeling them this way at almost twice the rate as white defendants." Further, the journalists found, "white defendants were mislabeled as low risk more often than black defendants."

There are risks in journalists venturing into complex, research-oriented territory. A team at Stanford has critiqued the methodology used in the "Machine Bias" story, for example.[18] But this is normal in the research world. The key is for journalists to express clearly any areas of uncertainty in their analysis and to condition their audiences to appreciate such uncertainty. Each day now, there seem to be new examples of bold journalistic experimentation and sophistication that matches forms of academic analysis and carries out broader data-oriented tasks that were once the sole domain of research institutions. Such activity is clustered around the higher-end, deeper-resourced organizations, but there is no reason that it cannot become more common. For such complex work to become more common, many more journalists will need to prepare for the field in different ways, to build the requisite skills and knowledge—and organizations will need to take different approaches to news.

The specific examples cited here of "deep dive" investigative, data-centered inquiry and reporting will always be, in some sense, special cases. No newsroom could dedicate those kinds of resources all of the time. Most stories will be daily stories. Obviously, journalism will not always be involved in heavy data collection or statistical analysis. The more regular benefit of greater knowledge in journalism will be its application in the judicious selection of what's important in the process of reporting on deadline. What remains clear from the major public reaction to such stories—and all of the important conversations they began—is that this new kind of deep knowledge can generate new networks of recognition that serve democratic needs.

Pictures in Our Heads

How does journalism contribute to general knowledge? And if practiced badly, how much might it foster general ignorance? Although simple on their faces, such questions contain layers upon layers of complexity. Scholars have been thinking about them for at least a century now. Revisiting two classic stories from the field's foundational texts can help us frame these issues.

The 1931 autobiography of Lincoln Steffens, one of journalism's great early muckrakers, contains a rightly famous passage that illustrates how journalism can profoundly shape our sense of reality. In a section titled "I Make a Crime Wave," Steffens recounts how, in an escapade driven by rivalry with his journalistic competitor and friend Jacob Riis, he decided to focus on more and more sensational crimes across New York City. Riis responded in kind. The triggering event was a familiar one to all beat reporters: both correspondents had "scooped" one another in sequence and were upbraided accordingly by their editors. Steffens and Riis then engaged in a contest of sorts, digging out all manner of crime to hype it. The public, and then-police commissioner (and future president) Teddy Roosevelt, suddenly found themselves deluged by what seemed an accelerating pattern of crime. Commissioner Roosevelt called in both reporters, asked them what in the world was happening, and told them to stop. The crime wave, such as it was according to Steffens' colorful account, then halted.[19]

The fact that news can change the public's perception of reality in dramatic ways may be obvious, but it is worth situating this point in deeper context: reality is always in some way mediated, whether by mass media, social media, or merely the people with whom we live and work. The Steffens incident highlights a more general phenomenon articulated by Lippmann in the opening passage of *Public Opinion*, in a section called "The World Outside and the Picture in Our Heads." He recounts a situation in which people of several European nations were living on an isolated island in the ocean when World War I broke out in 1914. According to Lippmann, for a long time the English, French, and German inhabitants of the island had no news of the outbreak of war. "For six strange weeks they had acted as if they were friends, when in fact they were enemies," he writes. Perceptions of reality are shaped by information—or, in this case, its absence. Lippmann goes on to make a more general insight: "The only feeling that anyone can have about an event he does not experience is the feeling aroused by his mental image of that event." These "pictures inside the heads" of people, "the pictures of themselves, of others, of their needs, purposes, and relationship, are their public opinions."[20] Lippmann's entire prescriptive and analytical project in the book becomes precisely how to facilitate a more accurate picture of the "world outside" in the minds of the citizenry.

It would seem that both the Steffens and Lippmann anecdotes may be artifacts of a predigital era; it is hard to imagine them unfolding in quite the same linear way, uninterrupted by other, competing information sources, in the era of email and Facebook. And yet new problems of misinformation, propaganda, fake news, and outright mainstream media neglect are as powerfully with us as ever. The basic problem of appropriately and proportionately mediating reality—how to condense and accurately represent the infinite data that constitute the physical and social world—continues in new forms, despite new technologies and platforms.

Knowledge and Journalism Practice

In recent years, a nascent literature has grown around the theoretical connections between knowledge and reporting. Patterson defines knowledge in this context as "systematic information," a category distinguishable from the anecdotal information gleaned from interviews and on-scene observations. Grasping these systematic patterns is a "key to devising accurate interpretations of what is observed or factually recorded." This capacity transcends tapping more online sources for information; rather, knowledge enables the "investigator to recognize things that would otherwise be misunderstood or go unnoticed." Patterson asserts that the Internet era, rather than making journalists less relevant, has created an even greater demand within democracy because the vast pools of available information are of such varying quality.[21]

Similarly, Wolfgang Donsbach notes that the "possibilities for journalists to do research and for people to have a voice—even in non-democratic systems—have never been better." Because of journalism's increasing "marginalization" amid a diversified communications landscape, it is necessary for journalism to redefine itself as a "knowledge profession," Donsbach argues, to make journalism "distinct again from other forms of communication—for the sake of the quality of the public discourse."[22] This idea of a professionalized knowledge-based role has also undergone further theoretical elaboration by Matthew Nisbet and Declan Fahy, who see the journalist variously serving public audiences as "knowledge broker," "dialogue broker," and "policy broker."[23] This can mean the translation of arcane research, the facilitation of critical conversations around complex issues, or the unpacking of thorny and difficult policy proposals that address societal problems. Mitchell Stephens has called for more "wisdom journalism," with media organizations staffed by "interpretive journalists." Journalists with expertise in specific areas might be organized "by subject matters akin to academic disciplines or subdisciplines"; they will be "looking not for news, but for the meanings and consequences of news."[24]

Several other trends help to contextualize these theories and new frameworks for journalistic practice. Kevin G. Barnhurst has noted a broad trend in journalism, dating back a half century, away from event-centered or "realist" reporting and toward "meaning-centered news" and "sensemaking."[25] This observation builds upon previous work on this issue by Barnhurst and Diana Mutz, in which they partly attribute the evolution toward meaning-centered news to a rise in quantitative data collection, enhanced computing capacity, an assimilation of social science approaches by journalists, rising education levels among journalists, and increasing professionalization.[26] Likewise, Katherine Fink and Schudson have documented an "enormous" industry shift toward "contextual" or analytical/explanatory journalism, a pattern that they call the "quantitatively most significant change in newspaper journalism between the 1950s and the early 2000s."[27]

The current journalism landscape has seen the rise of some newer outlets that have seized on new, knowledge-based capacities and trends, including deep "explanatory reporting" outlets and verticals such as Vox, FiveThirtyEight, the *New York Times*' Upshot, the *Washington Post*'s Wonkblog, and the data-journalism/investigative outlet ProPublica. The rise of data journalism is a kind of extension of the "precision journalism" first pioneered by Philip Meyer and others in the late 1960s and 1970s.[28] It was amplified by the movement toward computer-assisted reporting (CAR), which has empowered a certain class of journalists to begin executing what data journalist Steve Doig has called "social science done on deadline."[29] Cloud computing, sophisticated web-based software, an emphasis on learning computer programming, and increasingly accessible and creative data-visualization tools are also fueling the continuing convergence of social science and journalism in some respects. The programming language R, a go-to data science tool among academics, has been readily adopted by the likes of FiveThirtyEight and data journalism team members at the *New York Times*.

Journalists themselves appear to be cognizant of these larger patterns and their changing professional role in response. Media scholars Lars Willnat and David H. Weaver found in 2013 that 69 percent of respondents said that "analyzing complex problems" in society is "extremely important," the highest historical level recorded on the survey, which dates back to 1971. The scholars note that the level of response to that question is up an "astonishing 18 points from 2002" and that analyzing complex problems and investigating government claims are what journalists now believe are their two most important functions.[30]

Dewey long ago noted the need for the merging of timeliness and relevance with social scientific principles of inquiry. News, he noted, is definitionally a form of high alert for change and shocks to the continuous flow of events, and "the catastrophic, namely, crime, accident, family rows, personal clashes and conflicts, are the most obvious forms of breaches of continuity." The search for discontinuity is what makes news *news*:

So accustomed are we to this method of collecting, recording and presenting social changes, that it may well sound ridiculous to say that a genuine social science would manifest its reality in the daily press, while learned books and articles supply and polish tools of inquiry. But the inquiry which alone can furnish knowledge as a precondition of public judgments must be contemporary and quotidian. Even if social sciences as a specialized apparatus of inquiry were more advanced than they are, they would be comparatively impotent in the office of directing opinion on matters of concern to the public as long as they are remote from application in the daily and unremitting assembly and interpretation of "news." On the other hand, the tools of social inquiry will be clumsy as long as they are forged in places and under conditions remote from contemporary events.[31]

Dewey acknowledged that this synthesis of timeliness and depth will be no small task; in fact, he conceded there is a certain "ridiculous" quality to the suggestion. Yet the rise of available knowledge through the web and sophisticated tools that are instantly accessible online makes this vision much more achievable in our current moment. Tools of inquiry that are being developed through both data journalism and social media listening and analytics, as well as "learned books and articles" by the millions online, are now readily available. Journalists must seize these opportunities.

A Body of Knowledge

In an age of big data and networks, and of increasing polarization, journalism must get much closer to social science in its approach. As mentioned, this idea has long been advocated by the likes of Meyer.[32] Above all he advocates for more careful attention to journalistic method, which he believes could bring journalists closer to the goal of true objectivity. "Instead of implying that there is an equal amount of weight to be accorded every side, the objective investigator makes an effort to evaluate the competing viewpoints," he notes. "The methods of investigation keep the reporter from being misled by his or her own desires and prejudices."[33]

It is striking to compare journalism as a discipline to other fields, such as law, medicine, accounting, business management, and many more. "Almost alone among the professions, journalism is not rooted in a body of substantive knowledge," Patterson notes.[34] Here we might distinguish between content and process knowledge: *Content knowledge* means an understanding of a particular discipline, such as economics, health care, the environment, or criminal justice. By contrast, *process knowledge*, as Donsbach has defined it, involves awareness of issues of communication and potential bias and—at the higher, metacognitive level—the ability to evaluate the strengths and weaknesses of one's own reporting practices and of journalistic practices and strategies more generally. "If journalists know about, for instance, socio-psychological factors and group dynamics, they might resist more of the drives of 'pack journalism' and its often irrational decision-making," Donsbach notes. "If journalists know more about audience research, they [may] be able to present their messages in a way that might maximize not only attention to news but also, if employed in a responsible way, its cognitive processing by the audience."[35]

Although journalism may never be rooted in as specific a body of content knowledge as the fields of law, medicine, or other well-defined professions—journalism covers every field, after all—it nevertheless must train journalists to root their methods and practice more deeply in systematic information and data while also understanding theories and concepts that help put events and information in meaningful context. The locus of this debate, naturally, is journalism schools, in which the issue of how to prepare students for leadership in an industry with an uncertain future is the essential question.

Given societal and technological trends, the logical step for journalism is to imagine and delineate how preparation for the profession might better incorporate knowledge-related skills, interdisciplinary learning, and online engagement and news-gathering strategies based on deep knowledge of networks. G. Pascal Zachary writes that in an "era of pervasive digital networks that instantly deliver news with scant human help, the successful journalist will be, above all, a knowledge maker."[36] This echoes Donsbach's central idea of journalism as the "new knowledge profession." Such a transition is made all the more urgent by the fundamental challenge to journalism's core value proposition presented by the online environment. "Today, many journalistic functions have been stripped from the news media," Picard has noted. "Social media are now the primary centers of breaking news."[37]

Digital skills and fluency with tools are necessary, for sure, but they are not sufficient. Serena Carpenter suggests that as technology and the media business evolve, journalism students may be better served by having "adaptive knowledge" and more "theoretical knowledge" that is flexible enough to be applicable in new and unpredictable circumstances, which may make obsolete most skills with software and digital tools.[38]

The current debate joins a long-running discussion among journalism scholars. Michael Ryan and Les Switzer note that there has been a persistent tension within journalism education over balancing skills and *concepts*, or traditional academic content. "Many programs in the latter part of the 20th century evidently were oriented primarily toward skills development," they note.[39] In 2013, Jean Folkerts, John Maxwell Hamilton, and Nicholas Lemann—all former deans of major journalism schools—called for journalism programs to orient themselves more toward their universities, with faculty doing more academic research and with curricula engaging more systematically with the content-based offerings in other university disciplines. They note that earlier in the history of journalism education, some instructors were integrating many other disciplines from outside journalism, including a broad menu of liberal arts: "We see all three of [the] early strains in journalism education—practice-oriented, subject matter-oriented, and research-oriented—as essential."[40]

Similarly, G. Stuart Adam has recommended matching "elements of practice" to university disciplines, including "evidence-gathering and fact assessment" that has "authority not only in journalism itself, but in science, empirical social science (including statistical evidence), legal studies, and information science," in addition to "analytical and interpretative capacities nursed into existence through the formal study of ideas, on the one hand, and through specialization in the languages and forms of understanding marking a major discipline."[41] Finding connection points such as these across university curricula will be vital in strengthening journalism education along dimensions of knowledge.

New Knowledge Guideposts

As the discussion in this book has so far suggested, journalism education and training might focus on two key areas to help guide reform and the fostering of new competences and to chart new interdisciplinary pathways within universities. These might be loosely clustered under the categories of knowledge competences and network competences, although there are interconnections between the two areas; their separation here is merely for illustrative purposes.

Knowledge Competences

Journalists may cover so many subjects in the course of a career that a focus on a single area of specialization, particularly at the undergraduate level, may not be sufficient. The process of learning the fundamentals of a science-based or social science–based discipline, however, can expose a student to valuable analytical methods. Ultimately, the goal must be a form of training that prepares students to learn quickly and deeply in an online context, what information scientist Alison Head and I have called mastering the art of "knowledge in action."[42] The core competency, then, is a deep and flexible capacity to master complex issues—to acquire, so to speak, knowledge about how to use knowledge.[43] Under these broad competences, we might include specific items, such as the following:

• Literacy with basic statistics and strong quantitative reasoning skills. A knowledge of how research is conducted, including issues relating to sample sizes, confidence intervals and error bars, p-values, and common statistical analysis techniques such as regression.

• Awareness of how hypotheses are tested and theories are built. Understanding of the idea of an independent, dependent, and intervening variable that work to form an explanatory model.

• Skills in the manipulation and analysis of primary and secondary data in tabular form.

• The ability to map research discourses and to discern the rough "state of knowledge" within different disciplines and relating to certain questions.[44] An example might be the state of knowledge on an issue such as tough crime laws and deterrence, were a reporter to report on a policy debate in this area. This requires familiarity with online databases and a critical approach to the culture and conventions relating to how knowledge is produced and disseminated.

• Fluency with public records, the laws and procedures that allow access, and the typical norms and conventions of government agencies at all levels that determine how information is collected and synthesized.

• Web-based techniques for acquiring and presenting information, from scraping and accessing application programming interfaces (APIs) to visualizing data.

Of course, interviewing and gathering information in the field will remain core to journalism practice. We might imagine these new competences as being woven together with the traditional tasks, creating an iterative process (see figure 7.1).[45] Any provisional model of an "ideal" journalistic process will not be appropriate in all situations; journalism is too varied for any single model. However, we might stipulate that it should begin with identifying existing assumptions and biases—a first step for journalists striving toward true learning and impartiality—while bookending the process with transparency, both in terms of showing one's work and being clear about the limitations and uncertainties in the reporting so far.

By grounding their practice in knowledge and then expanding the public's insights through reporting in context, journalists are at lower risk of committing the litany of lamentable practices well known in media criticism: the personalization of issues; the thin "he said, she said" stories; the political game and horse race narrative; or the "churnalism" of the retyped press release. These lamentable patterns embedded in coverage have increasingly divorced journalism from democracy and alienated citizens from news media. In this regard, Patterson has observed: "When reporters must file quickly, without the opportunity to observe or conduct interviews, they

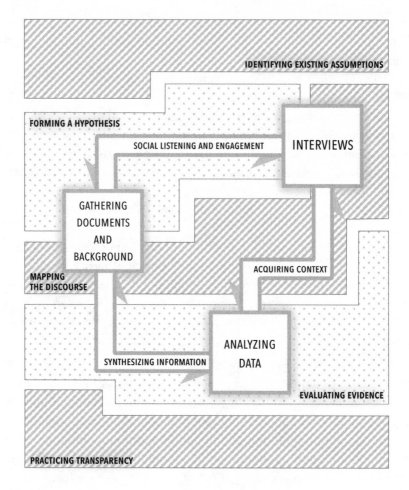

Figure 7.1
Guiding principles and patterns of workflow for journalism in an age of networks.
Source: Steven Braun and John Wihbey.

have no place to turn except to what they already know. Knowledge is the best remedy for hastily concocted, wrongheaded story lines."[46]

Skills with Networks

A review of much of the current professional discourse, the journalism studies literature, and the instructional material read in courses suggests that the discipline has not become fluent with a new and broad set of scientific disciplines that have direct bearing on news creation and dissemi-nation—specifically, network science and theory. Social network analysis and areas such as graph theory and a wide, related range of data science techniques have become the basis for the great modern information com-panies such as Facebook and Google. There is real potential in journalism drawing on network-related fields, or at the very least better understanding their language and frameworks. This knowledge can help journalists under-stand audiences, sources, and opportunities to connect groups and nodes of information, sparking networks of recognition.

Network Competences

Social media management and some basic analytics courses are being taught at some colleges and universities, but it remains the case that jour-nalism schools have not to date capitalized on the revolution taking place in the network sciences and computational social science. Competence in this area would extend well beyond the ability to disseminate information to wider audiences; it would also relate to the analysis of online publics and information spaces to help fuel deliberative discourse and to analyze patterns of sentiment, behavior, and discourse. Understanding networks is also key to investigating the sources of online misinformation. The network competences might include the following:

• A rich understanding of how the web operates at a technical and behavioral-social level; of the basic patterns of information flows; and of the archetypal forms—tight crowds, community clusters, hub-and-spoke networks, and so on—that online communities commonly take as dis-course unfolds.

• Comprehension of website and social media analytics, as well as the meanings of, and debates over, the proper way to measure success for jour-nalists in an online environment.

• The ability to map, using social network analysis techniques, the shape, sentiment, and size of online communities.

• An understanding from business and cultural perspectives of how com-mercial, third-party online platforms operate and are governed and how form and function may affect information flows.

• A theoretical sense of socially networked behavior, informed particu-larly by the quantitative social science and network science literature and including the techniques of social network analysis.

Journalists cannot be expected to be network scientists. Some network-related work can be achieved more or less easily as a function of quality software development. Experts who specialize in the analysis of networks often admit that, so far, the tools used for network visualization and analysis are not as intuitive and well-designed as the tools for more general data analysis and visualization purposes.

Network analysis techniques are not always used for the purposes of audience research or directed solely at social graphs of persons online. The computational techniques used in network analysis might see their most powerful use as an aid in mapping the relationships among persons and entities in public records. Data journalist Jonathan Stray analyzed thirty-four media stories in which network analysis of some sort was used; indeed, looking for relationships in public documents is one primary use of network analysis, although visualization is also frequently performed. "Network analysis has now been in use in journalism for several decades, and is increasing in popularity," he notes. "But the journalistic meaning of the term is sometimes different from what is meant in computer science or sociology."[47]

Stray points to several examples that represent the current use of these techniques: a *Seattle Times* article that mapped figures in the local art scene; a *Tampa Bay Times* investigation of the relationship among car thieves; and the Pulitzer-winning investigation run by the International Consortium of Investigative Journalists (ICIJ), which built a database to allow a coalition of some three hundred reporters globally to find relationships hidden in the millions of leaked documents that showed corruption in the global banking system.[48] "Perhaps the journalist's use of network data is closest in spirit to the 'link analysis' of criminal investigations," Stray observes. "At the frontier of investigative journalism, graph databases are emerging as the representation of choice for fusing large and complex data sets."[49] New computational techniques and software must be developed to harness the possibilities of network analysis.

Finally, it should be noted that the art of *narrating networks*—journalistic visualization techniques and typologies that represent for audiences interconnected persons and data points—is still a field in its infancy. The use of network imagery in visual journalism is only just now coming into wider use; how well these forms and images are able to connect with and engage audiences deserves further study.[50]

Journalists' Attitudes and Capabilities

If we grant that journalists must achieve this greater fluency with social science and data and apply this knowledge in their practice, it is worth asking how far away the field currently is from achieving any such vision, both in terms of the overall, current state of competences in this regard and the state of academic training. Where does the field stand?

Three consecutive annual surveys (2015–2017), all of which I conducted through the Shorenstein Center on Media, Politics and Public Policy, shed important light on issues of current capabilities and attitudes toward new skills. The most extensive survey illuminates attitudes and self-reported capacities relating to research and data across a wide range of measures.[51] The survey respondents included 1,118 full-time working journalists and 403 journalism educators.[52] The survey was sent to roughly nine thousand identified journalists and educators. The underlying demographic characteristics of the respondents roughly conformed to the news media business at large, in terms of age, education, employment, tenure, medium, and gender.

In keeping with long-standing tradition, journalists reported that conducting interviews, both on the phone and in person, in addition to on-location reporting, were the information-gathering methods that they most frequently use. However, the practice of drawing on and using government data and research studies also figured prominently among the methods used, with about half of all journalists surveyed saying they use these methods frequently. Still, there is clearly room for improvement in terms of the methods with which research is accessed: The most frequent way that journalists report drawing on research for stories is through following links in other articles and calling up experts. Only about one-fifth reported frequently using Google Scholar—by far the most common tool used by researchers themselves to access scholarly literature of all kinds—and other academic databases saw little use at all.

Reporters and editors are, in other words, mostly reliant on other sources to locate research relevant to their stories. This lack of direct access to research not only indicates a certain passivity on the part of journalists but also highlights their susceptibility to biased interests—even individual researchers or research institutions—who may carry a hidden agenda. As

Patterson notes, "Journalists' knowledge deficiency is a reason they are vulnerable to manipulation by their sources."[53]

What was striking to see among journalists in that same survey, however, is a general enthusiasm for the possibilities of research: three-quarters said academic research could be "very helpful" to journalists in terms of deepening story context and strengthening story accuracy, and nearly 70 percent said such research could be very helpful in countering misleading claims by sources. The value of systematic information and knowledge to journalism seems clear enough to the field's practitioners, at least in this sample.

Further, a statistical analysis on the 2015 survey data found some important correlations. Journalists producing stories for a national audience, versus a regional or local audience, were more likely to report using research studies in their story over the past year and to report speaking frequently with academic experts. This might be expected, given that national stories may look for wider, more general context. Educational attainment was associated at a significant level with greater frequency in speaking with academic experts, although having advanced degrees was not associated at a significant level with using research studies more frequently in stories. Further, among journalists who reported having enough training to perform statistical analysis on their own, there was a significant relationship with both using more research studies in stories and greater frequency of contact with experts. As might be expected, journalists covering science-heavy beats, particularly in medical science, were more likely to draw on research studies in their stories and speak with experts more frequently. Reporters covering politics were less likely to say they drew on research studies as compared to those covering other beats, although they were just as likely to be in touch with experts. Among different news mediums, journalists in television/video were less likely to draw on research.

One of the chief takeaways from the survey was the striking disconnect between journalists' self-reported skills with various data-related tasks and the amount of value they know such skills have for journalism. News media members know they are not as prepared as they should be (see figure 7.2). In terms of data skills, only about one in ten journalists in the survey said they were "very well equipped" to perform statistical analysis on their own, and 46 percent said they were "somewhat" well equipped on that skill measure. However, troublingly, only one-quarter of journalists said they were very well equipped to interpret statistics generated by other sources, with

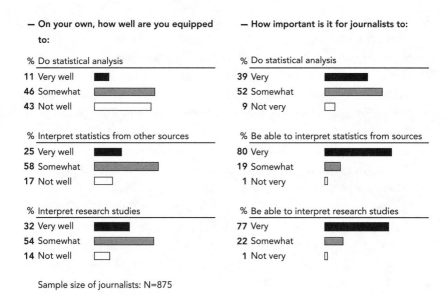

— On your own, how well are you equipped to:

% Do statistical analysis
- **11** Very well
- **46** Somewhat
- **43** Not well

% Interpret statistics from other sources
- **25** Very well
- **58** Somewhat
- **17** Not well

% Interpret research studies
- **32** Very well
- **54** Somewhat
- **14** Not well

— How important is it for journalists to:

% Do statistical analysis
- **39** Very
- **52** Somewhat
- **9** Not very

% Be able to interpret statistics from sources
- **80** Very
- **19** Somewhat
- **1** Not very

% Be able to interpret research studies
- **77** Very
- **22** Somewhat
- **1** Not very

Sample size of journalists: N=875

Figure 7.2
How journalists rate their own data abilities and see data's importance.
Source: 2015 survey, Journalist's Resource, Shorenstein Center, Harvard University.

58 percent saying they were somewhat equipped. In terms of their self-assessed ability to interpret research studies, about one-third said they rated their ability very highly.

In the 2016 Shorenstein Center survey (of about seven hundred full-time or nearly full-time journalist respondents), journalists were asked about the barriers to their drawing on more research in their reporting. Some 60 percent pointed to a lack of access, such as online paywalls; a third said they didn't know enough about research methods to assess accuracy; and nearly half were worried that potential bias and funding sources were not clear and therefore could not be assessed.

Another survey conducted that same year asked a group of 110 health science journalists about their views on dealing with the academic research and scientific institutional worlds. The publication of new scientific studies, they said, was the lifeblood of their reporting cycles, as nearly half indicated that their stories were frequently triggered by new publications; nearly a third of journalists said government reports were a frequent trigger for stories too. That said, about 40 percent said having inadequate time inhibited their ability to produce quality stories in a knowledgeable way, one-third cited difficulty in assessing the validity of new findings as a major problem, and another one-third blamed the need to frame stories in a way that will catch the public's attention. About one-third of health science journalists said insufficient space for their stories was a major barrier to quality journalism. On the question of why health and medical information gets misrepresented to the public, about two-thirds of journalists surveyed blamed industry-funded scientists either "a lot" or "somewhat"; a roughly equivalent number of journalists blamed nonprofessional sources of news, such as blogs or talk shows.

The 403 persons in the educators group represented in the 2015 survey hailed from a diverse mix of schools and backgrounds. In the sample, about half were at research universities and half at teaching-oriented or liberal arts institutions (a few percent were at community or technical colleges). In terms of their primary preparation for work, 46 percent cited having been a professional journalist, 15 percent cited an advanced academic degree, 30 percent cited a mix of professional journalism and a training degree, and 8 percent cited a communication background.

It is interesting to note that the journalism educators consistently did not value academic research for the practice of journalism as much as the working journalists did. This may be explained by a variety of factors, including that many current instructors may have left the practice of journalism before the digital revolution provided so much instant access to the research world, giving rise to the likes of Vox, FiveThirtyEight, ProPublica, and the Upshot at the *New York Times*. Further, although journalism academics had greater self-reported levels of fluency with data and statistics in the 2015 survey, it was still the case that only 45 percent of journalism professors said they were very well equipped to interpret statistics from other sources, and only about one in five said they were very well equipped to perform statistical analysis themselves. Only 5 percent of journalism academics said the academic program in which they teach provides "extensive" training in statistics to students (43 percent said there was "some" preparation). Further, just 8 percent said there was extensive training in interpreting numbers from sources, although 62 percent said there was some such preparation for students.

Overall, the evidence from these surveys speaks to the need for significant capacity building among working journalists, educators, and students.

It also speaks to the need for more scholarship in this area as there has been little other empirical research with respect to how much journalists draw on knowledge. Other research investigations of specific reporting areas are not heartening. For example, a 2015 study by Holly Yettick found that, in contrast to science or medical journalism, education journalists almost never cite peer-reviewed research. This means that the findings of the most reliable source of information about education are "barely a blip on the radar of American education reporting."[54] Studies of basic media accuracy also broadly speak to this problem: one study of American newspapers found that 61 percent of articles contained errors.[55] The public has consistently complained about a lack of knowledge on the part of journalists on basic economic, legal, and government issues they are covering; public evaluations of news accuracy continue to decline.[56]

If we accept the obvious fact that the world is becoming increasingly data-driven and knowledge-focused, it would make a great deal of sense for journalists to prepare themselves better for this environment. Further, to maximize their contribution to the functioning of democracy and provide the public with relevant and accurate material, it is inarguably essential that journalists, as a whole, become much better equipped to make selective judgments about the importance of issues and information and parse complex quantitative patterns amid the blur and noise of the online public sphere. It is unrealistic to expect all in journalism to become statisticians, and certainly not every beat and subject requires deep familiarity with the relevant academic literature. Still, we should value these skills and, as an aspiration, look to impart them as pillars in media work, as vital parts of the body of knowledge that constitutes the evolving discipline in the digital era.

Characterizing Uncertainty in News

A greater embrace of knowledge also means a greater commitment to capturing uncertainty. What the "boundary-free" pattern of knowledge in the digital ecosystem requires of journalists may be a much more decisive emphasis on transparency relating to information-gathering and interpretation, as well as a humbler approach to the verdicts rendered and the pictures presented to the public. The public has little understanding of the methods that journalists employ or the codes of ethics to which journalists adhere. An even wider disjunction between trends in the world in general—in which a lack of hierarchy and endless modification and interpretation prevail across the web—and journalism—in which purportedly authoritative statements about the world are rendered in final form—threatens to undermine the reporting profession even more. Surely it is one of the reasons that trust in the press has declined so precipitously across the United States.

If there is a general trend that we might cite as particularly relevant in this regard, it is the broad, secular decline in terms of trust in US institutions of almost all kinds—from Congress to religious institutions, from the Supreme Court to news media, which have seen among the sharpest declines in public evaluations of trust.[57] It is worth thinking about a journalistic response to our age of anti-institutionalism.

We may be entering an era in which there is simply less tolerance for certainty and definitive judgments on controversial issues of any kind—and news media, having hit near rock-bottom in terms of general trust, should look more explicitly to state uncertainty and lack of information up front. Strategies of intellectual transparency may aid the creation of networks of recognition, insofar as citizens may connect with the network for differing reasons, drawing on different civic epistemologies and modes of understanding. Stories and narratives that express areas of uncertainty—where further knowledge is needed—may be less likely to alienate and more likely to draw in diverse publics for varying reasons.

Some observers and scholars have also called for a new journalistic ethics based on transparency itself, replacing older ideas about objectivity.[58] All of these changes and trends speak to the need for journalists to frame their work more as an ongoing process than as a finished product. It also speaks to the need to make clear notions of uncertainty and fallibility. A world of wide, populist, and unsettled knowledge also means changing audience expectations and greater accountability (and punishment) for errors of hubris.

Nate Silver, the data journalist who founded FiveThirtyEight, notes that news media often fabricate clear trends based on noisy data, creating what are pejoratively called *media narratives*. Media have a "probability problem," in his view, stemming in part from lamentable professional incentives. "One can understand why news organizations find 'the narrative' so tempting," Silver states. "The world is a complicated place, and journalists are expected to write authoritatively about it under deadline pressure."[59] Silver's prime case studies in media probability problems relate to political polling and elections, as well as extreme weather events such as hurricanes—classic cases in which journalists frequently overinterpret forecasting.

Silver's point can be applied more generally. The need for analytical humility and the clear communication of the possibility of error, or future deviation from current trends, is vital across all beats. Journalists must find ways to acknowledge uncertainty and to modify interpretations as the world of increasingly "unsettled" knowledge demands it. This issue of characterizing uncertainty, anathema to a media world of blaring headlines and confident statements delivered by talking heads and pundits, is a crucial issue, one in need of a great deal more examination and study.

Reporters often have built their picture of what they believed was generally true through anecdote, through observation and interview. Surely such tools always will be useful for exploring human experience and events. Data analysis cannot always capture the texture and nuance of lived experience, but being able to place anecdotes in the context of systematic information is vital if the journalistic picture of the world presented to the public is to be reasonably accurate.

Histories of Uncertainty

Let's consider the lack of trust in media from a historical perspective. On June 11, 1807, President Thomas Jefferson wrote a letter in reply to John Norvell, a young man seeking career advice and asking Jefferson's opinion on becoming a newspaper publisher. Earlier in his career, Jefferson is known to have had a few positive and memorable things to say about newspapers. In a 1787 letter, he famously wrote that "were it left to me to decide whether we should have a government without newspapers or newspapers without a government, I should not hesitate a moment to prefer the latter."[60] However, having been the object of sustained criticism by certain quarters of the press throughout his presidency, Jefferson was not so sanguine two decades later. In a letter dripping with sarcasm, he told the young Norvell (a Kentucky native who would go on to edit periodicals in Lexington, Kentucky; Baltimore; and Philadelphia)[61] that, if he were to start a newspaper, the whole endeavor should be wholly and strictly reformulated:

Perhaps an editor might begin a reformation in some such way as this. divide his paper into 4. chapters, heading the 1st. Truths. 2d. Probabilities. 3d. Possibilities. 4th. Lies. the 1st. chapter would be very short, as it would contain little more than authentic papers, and information from such sources as the editor would be willing to risk his own reputation for their truth. the 2d. would contain what, from a mature consideration of all circumstances, his judgment should conclude to be probably true. this however should rather contain too little than too much. the 3d. & 4th. should be professedly for those readers who would rather have lies for their money than the blank paper they would occupy.[62]

Jefferson's acid take on the press—the chapter on "truths" would be "very short"—and his evident exasperation is no doubt a reaction to the river of negative ink he saw as president, much of it based on innuendo, gossip, and hearsay. In 2017, the *Washington Post* noted that Jefferson stands as a kind of progenitor to President Donald J. Trump, who likewise has shown absolutely no affection for journalism.[63] (Both, notably, were favorably disposed to the press earlier in their careers.) Jefferson's views also provide further perspective on the "fake news" crisis of our current political moment: more than two centuries prior to the 2016 presidential election and its fallout, the sitting president of the United States noted that "nothing can now be believed which is seen in a newspaper. Truth itself becomes suspicious by being put into that polluted vehicle."[64]

The opinions of politicians aside, what is more interesting is taking seriously Jefferson's idea that news outlets might more explicitly label certain representations as "probabilities" and "possibilities." The issue of formally expressing probabilities has remained a hot one in science journalism, in which the task of reporting typically involves accurately capturing researchers' own expressions of uncertainty in a particular study. The communication and journalism studies literature has examined this issue fairly extensively, and some research has suggested that more nuanced, or hedged, portrayals of scientific research can engender more trust with news consumers.[65] Whole books have been written about how journalists can better handle numbers in general, including quantification of uncertainty; it has long been a pet peeve of researchers that journalists regularly distort numerical information or get it wrong.[66]

In some ways, however, science journalism, or journalism specifically about any kind of research, is a special case, in which the expression of uncertainty is already baked into the subject. In such cases, journalists need to do a better job of reflecting any probabilities articulated in the given research paper and avoid the temptation to overinterpret or sensationalize. Quality interviews with the researchers involved and careful attention to nuance are the key in these cases. There also should be due attention to critical views of other scientists not connected to the research in question and even of researchers not totally embedded in that particular, narrow corner of science, as statistician Andrew Gelman recommends. "If journalists go slightly outside the loop—for example, asking a cognitive psychologist to comment on the work of a social psychologist, or asking a computer scientist for views on the work of a statistician—they have a chance to get a broader view," he suggests. "To put it another way: some of the problems of hyped science arise from the narrowness of subfields, but you can take advantage of this by moving to a neighboring subfield to get an enhanced perspective." Gelman advocates that journalists bring an attitude of skepticism to new findings as a form of postpublication review, "conveying to readers a sense of uncertainty, which is central to the scientific process."[67]

In terms of graphical representation of uncertainty in data, journalism has a less-than-stellar track record, even among the most sophisticated and elite outlets. Amanda Cox, editor of the Upshot at the *New York Times*, has said that she can find only eight examples in the paper's graphics archives in which journalists "formally expressed some type of confidence interval." In the instances in which uncertainty was articulated in this way, she says, it generally appears that the *Times* was somewhat "forced" to do it to account for an odd or marginal trend. An example she gives is a story in which child obesity was essentially plateauing in the United States, but plotting the data made it look like there was a slight increase. "I do think that if we got more comfortable with uncertainty," Cox said, "if we got more comfortable with the fact that we don't know the future but that we can have educated guesses about things, I think that there are real world implications and policy consequences to that."[68] The real-world implications may be that audiences both benefit from and appreciate the increasing honesty and humility from news organizations.

The lack of formal expressions of uncertainty in journalism may not stem from laziness or malpractice, but rather from a reasonably well-founded belief that audiences simply will not understand. Alberto Cairo has pointed out that the "cone of uncertainty" that is often used to represent the path of hurricanes on weather maps is frequently misinterpreted by viewers as the size of the actual hurricane. Cairo hypothesizes that there are likely historical reasons that visual representations for public audiences do not include expressions of uncertainty: most of the classical and canonical forms (e.g., scatter plots and bar graphs) for visualizing data were created at the end of the nineteenth century, before the advanced science of uncertainty and sophisticated statistical techniques and data analysis were developed in the twentieth century (by the likes of Ronald Fisher and John Tukey, among others).[69] It remains the case that the public has trouble interpreting simple plots showing the correlations between two variables, never mind more complex visualizations that highlight uncertainty.[70] The research on how lay audiences respond to visual manifestations of uncertainty is not yet well developed.[71]

For stories in which facts and knowledge are in dispute, there may be good reason for journalists to show as much transparency and uncertainty as possible. Damaris Colhoun has noted in *Columbia Journalism Review* that "uncertainty can be a powerful tool. When reporters embrace how little they know, resist forming conclusions, and share their doubts with their readers in a form that breaks with convention, they may wind up getting closer to the truth." This speaks to a more general point: the reporting of nearly any short-term statistical trend—rising violent crime, more car accidents, increased home valuations, fewer drug overdoses—more frequently should be reported with substantial caveats. Trends fluctuate for all sorts of reasons; many short-term patterns are just random noise and eventually regress to a historical mean.

Technical questions over expressing uncertainty in the context of science, research, and data visualization should also prompt us to think more broadly about how to convey our general confidence in any conclusion based on the available evidence, across stories that may involve interpretations relating to specific events—things that are episodic in nature. What should we make of a local election result or a year of lower test scores in a school or a team's streak of victories? Some issues are so mundane that it might seem pure nerdy overkill to assign some level of confidence to a story, yet many might benefit from a journalist, or a group of journalists and editors, saying in effect, "We believe X explains Y with high/moderate/low confidence, based on the following evidence." This would be to take a page from, for example, the intelligence community, which is always working with incomplete information to render defensible judgments.

Charles Weiss has proposed using legal terminology to express scientific uncertainty, a body of language and concepts that are more familiar to the lay public. He has proposed using legal-inspired language over a 0–10 confidence interval scale delineated by terms such as *beyond a reasonable doubt, preponderance of evidence, probable cause*, and *no reasonable grounds for suspicion*. The scale runs from extremes of "impossible" to "beyond all doubt."[72] Whether or not this precise language would work in journalism is unclear, but it points to an important problem that would need to be solved. Fact-checking organizations, such as PolitiFact, currently use sliding scales to evaluate truth claims, and other such norms could evolve.

It might be a reasonable goal for news outlets to create a common language that can better convey degrees of certainty to audiences. Implicit in this is a need for journalists to better understand how hypotheses are tested and theories built and to bring a more scientific framework to the evaluation of evidence.[73] As discussed, some of the research evidence suggests that audiences would appreciate this; it is possible that it could even depolarize audiences a bit on controversial issues, putting news media more in the role of explicitly fallible referee rather than judge and executioner. The other necessary piece of any such efforts, as mentioned, is transparency. "Journalism is not scholarship and does not generally use bibliographies or footnotes," Nicholas Lemann notes, "but you should use attribution in your work in such a way that readers and colleagues can see, to the greatest extent possible, where your information came from and how you have reached your conclusions."[74] Showing the data and the documents, the body of evidence, is becoming a new norm in many forms of journalism.

In an age of increasing public distrust and skepticism, reframing the journalistic product by more clearly approaching and articulating stories as hypotheses that can be tested and that are susceptible to further evidence is an essential move. Paradoxically, an era that demands greater knowledge and closer proximity to systematic data from journalists also requires them to be less certain. But stating one's methods and assumptions, and staying open to new evidence, is the very core of science. Journalism will never be science, but it can both get closer to reality and potentially better engage audiences through an approach that emphasizes depth, self-awareness, and humility. Articulating limits, unknowns, and uncertainty may also invite public participation, potentially creating more space for diverse networks to discuss and share information.

8   Questions for Engaged Journalism

Engagement can be a nebulous concept. Although engaging audiences can sound like an unambiguous good, journalism is very much still working out what its role and identity might be in this new terrain. There are a number of complexities from a theoretical, practical, and ethical perspective that we must acknowledge—and that journalism will undoubtedly grapple, indeed struggle, with in the years ahead. Let us, then, consider briefly some of the internal debates within journalism over direct engagement with audiences. New ideas and research continue to be produced in this area, but I want to frame the debate in its broad outlines.

Fostering networks of recognition can be achieved through many forms of journalistic activity. The key, as discussed, is to discern and articulate citizens' stakes in issues, to engage them, to create incentives to learn, and to facilitate collective sense-making and public knowledge. For journalism to retain its value, it cannot become merely one more voice barking in the sea of social media content. The ability to apply knowledge to issues and problems and channel attention toward important civic concerns accordingly gives journalism its distinctiveness. Often missed in discussions of engaged media work is this need for greater knowledge. If journalists want to convene discussion with audiences, that is a good thing. If they have nothing to say themselves, no authoritative or well-researched context to add, then journalism-facilitated dialogue may make little progress.

There remains a philosophical debate in journalism circles about how deep community engagement and collaboration with potential audiences and sources should run. Many news outlets have implemented systematic engagement practices, and their social media desks—still often siloed from core reporting processes—almost uniformly try to promote stories to influencers. Hard questions remain: How much is journalism currently prepared to engage with the world of social networks in a constructive and impact-ful way? And how much is journalism prepared to hand over some agenda setting to the digital crowd? How much *should* it? Certainly, teams of social media managers and editors have been hired and deployed by every major news organization. Yet exactly what their mission should be, and how all other members of newsrooms should be leveraging and participating on social media, remains unclear.

As Jake Batsell notes, there continues to be no agreement either by aca-demics or practitioners on exactly what constitutes "engagement." Many focus on the concept of listening, while others mention audience members taking action of some sort prompted by the journalistic content, whether online sharing or offline conversation. Batsell offers five guiding principles for audience engagement by news organizations: the convening of audi-ences in person, interaction with audiences at every step, a focus on niche audiences that are passionate about issues, empowering audiences to satisfy curiosity, and measurement of success with a view toward both editorial and business goals.[1] Many of these principles are now widely employed by news organizations, particularly digital startups and the more forward-thinking traditional news outlets.

My Northeastern University colleague Dan Kennedy has analyzed these dynamics in the context of local news and concluded that news outlets increasingly must play a convening role, with journalism serving as a hub of community dialogue. Community news practice provides a forum, a spark for engagement, and a "gathering place." Older notions of journal-istic outlets just "serving the public" are "no longer enough. Rather, the public they serve must first be assembled—and given a voice."[2]

Although many novel questions about journalists' relationship with audiences are being raised in the social media era, key ideas that frame this area were developed long ago. In 1975, University of Chicago sociolo-gist Morris Janowitz helped bring the notion of the journalist as profes-sional gatekeeper into currency.[3] The fields of media criticism and critical communication studies have used this concept to explore how journalists construct reality for the public, selecting certain facts while disregarding others, imposing value judgments, and simplifying events into reductionist narratives. Journalists have been criticized, often based on strong evidence, for exercising gatekeeping power in the service of excluding marginalized groups or validating existing power structures.[4]

Because of these dynamics and more, over subsequent decades influential academics such as Jay Rosen have been encouraging a deep rethinking of how journalists relate to the public—embodied in the public or civic journalism movements in the 1990s—and how the "people formerly known as the audience" are playing a new role in the production and circulation of news and information.[5] As Bill Kovach and Tom Rosenstiel note in their influential books *The Elements of Journalism* and *Blur*, the gatekeeper metaphor masks multifaceted roles that journalists play in relation to the public.[6] Traditionally, these have included the roles of authenticator, sense maker, bearing witness, and watchdog. Further—and relevant to the new possibilities enabled by the online world—there are myriad other functions: curator, or intelligent aggregator; forum leader; empowerer; role model; and community builder. Engaging audiences can take on many forms, with gatekeeping and hierarchical power dynamics at work in various degrees along a continuous spectrum.

The ICTs and platforms that allow citizens to broadcast and share their own perspectives should, in theory, make journalism institutions more responsive to broader public concerns. However, there continues to be a basic tension that has begun to frame conversations across journalistic professional groups, conferences, and academic studies: Will this new audience data serve to help inform the news product, or will it merely be used to, in effect, spread and sell the news product? Put another way, who inside a news organization "owns" the audience—the editorial side of news institutions, or the business side?[7] Or is this binary a false dichotomy? Do news organizations need to reconceive their role radically in relation to audiences? Newsrooms that have tried to implement such notions of engagement have shown varying degrees of success in the Internet era; convincing journalists themselves to reorient their practice toward community deliberation has sometimes been difficult.[8]

There are many active debates on how newsrooms should be using, for example, social metrics.[9] Not everyone agrees on the degree to which audience engagement and cocreation should occur. "We want to challenge our readers, and we want them to trust us to do that—but we can only do it if we stay one step away from what they are saying," Adam Smith, the deputy community editor of the *Economist*, writes. "This approach is not self-important; it is a relentless attempt at delivering on what our subscribers ask of us. They want us to challenge and inform them; we cannot do that if we are forever reacting to what they are saying, publishing stories based on what they are consuming, and listening in to the ensuing discussion."[10]

Many believe that the key emerging shift is seeing social media less as either a distribution channel or even another sourcing opportunity, but more as a map of community needs, policy-related confusions, and open questions that demand a journalistic response. This requires a substantively different mentality than that which informs pure "click chasing," a practice that has characterized certain forms of social media monitoring and engagement by some media organizations—and attracted a fair amount of criticism and reflection within the industry.[11] A more promising direction may be to employ the concept of *impact*, defined flexibly by news organizations to fit their mission, to help delineate notions of success or failure.[12]

Often, the definitions of engagement unfortunately leave its measurement either intellectually impoverished (e.g., clicks, pageviews) or a bit vague (e.g., "empowering audiences"). In this regard, journalism might look to borrow and draw on the definition of engagement proposed by network theorists such as Alex "Sandy" Pentland of MIT. In his important book *Social Physics* (2014), which details new insights and experiments in social influence and collective intelligence, Pentland defines engagement in terms of social learning in a group, or "the process in which the ongoing network of exchanges between people changes their behavior."[13] This definition goes well beyond anything currently operationalized in journalism. Journalism institutions tend to not care about measuring impact in this kind of way, although this is beginning to change. Some organizations have piloted user surveys and other learning measurement tools. Thinking more in these terms of civic learning might help provide a richer way of conceptualizing what the purpose of journalism ultimately is.

What is clear, however, is that current ways of measuring success seldom take into account measures that we would recognize as contributions to democracy or an informed public. "Even the best editorial analytics," Federica Cherubini and Rasmus Kleis Nielsen note, "continue to be constrained by the difficulties involved in defining and measuring many of the things that news organizations aim to achieve and is beset by a whole range of data-quality and data-access issues, exacerbated by rapid changes in the media environment."[14] Newsrooms might use samples of their subscriber and member bases to test learning through small surveys, which could be complemented by web analytics, such as time spent with stories, as well as engagement measured through links and social media.

Engagement and learning can also be facilitated by practical measures, such as improving website interfaces, speeding up page loading time, and creating social media tools and comment spaces that invite audiences to become active participants and cocreators. The Engaging News Project (Center for Media Engagement), run by Natalie "Talia" Jomini Stroud and her team at the University of Texas at Austin, has made important insights in this regard. That project has continued to advocate testing new strategies and applications with audiences to keep up with the fast-changing expectations of digital audiences. "We are only beginning to understand why people share news content," Stroud notes. "It seems likely that what makes content more or less shareable differs by topic. Some topics may require a compelling visual to increase sharing, others a catchy headline. I also suspect that what 'works' changes over time as people grow tired of a formula."[15]

A Science of Engaging Content?

The idea of robust engagement is related to the concept of virality, insofar as contagious content suggests that publishers have hit the right note for getting the public's attention. Predicting robust engagement and virality are not identical, but they are interrelated. Sharing on social media is a chief measurement of a story's capacity to engage, after all. Many outlets have adopted practices such as *A/B testing*—in which a sample of the audience is shown multiple headlines, pictures, and story leads—to monitor patterns of performance for a given story.

For more than a decade now, newsrooms and content marketers of all kinds have been trying to optimize obvious factors, such as the best time of day to post a story on social media. Much of the online world has tempered expectations, pivoting away from early delusions of viral grandeur to the more modest position of being just vigilantly data-driven—and acknowledging that although perfection is not achievable, marginal gains in web traffic and attention are within their grasp.

Yet the digital holy grail—the dream of a true viral media science—remains glittering on the horizon for many in the publishing industry. In the spring of 2017, I attended a conference in New York City that gathered many of the millennial-centric news outlets most famous for their incredible social media success, from VICE News and NowThis to BuzzFeed and alums of Huffington Post and Facebook. Organized by NewsWhip and held at the "luxury boutique" Bowery Hotel in the East Village, the gathering was both hardheaded and idealistic in its themes and spirit, with digital publishers of all kinds acknowledging that achieving breakthrough virality is no easy thing—and yet devoting tremendous time and resources toward improving their outlets' reach through social pathways.

NewsWhip CEO Paul Quigley channeled the collective desires of the hundreds of publishers gathered by sounding the day's theme, a project to pin down the "science of content." Given "infinite competition for mindshare," there are pressing reasons for publishers to become more scientific in their targeting of audiences. The digital world becomes busier every day, even as humans' capacity to focus attention remains constrained by the 168 hours in a given week.

As Quigley and others noted, we still do not really understand why ideas spread, but in the coming years publishers will continue to try to unlock the "DNA" of stories. "We never really had evidence before about why some stories work and others do not," he said. "Now, we can analyze social data and see the patterns. Successful stories are built on analytics." Given this new scientific knowledge, though, there will be tensions—namely, between where the audience is pulling an outlet and the outlet's "integrity," as Quigley put it.

Many news organizations and analytics firms that are helping to fuel the social revolution in online publishing have had to grapple with the trade-off between pushing content that will see a lot of clicks and maintaining standards of quality. It's a tricky balance between short-term gains and potentially losing brand identity—and soul. Several years ago, Tony Haile—the one-time head of Chartbeat, an analytics company widely adopted by publishers—began hammering away at the theme of needing to "change the metric," to value something other than raw pageviews or unique visitors.[16] His solution? "Engaged time," which combines various measures to produce a combined index score. Others have developed similar measurement concepts, such as attention minutes, favored by the viral content publisher Upworthy.[17]

As Rosenstiel notes, however, news publishers in particular need to think about each of their topics of coverage in more differentiated ways to overcome the problem of "terrible analytics." He states that certain "topics are always going to be inherently more popular than others. ... By breaking down coverage by topic, publishers can begin to avoid the trap of clickbait—trying to make every story a breakout hit. The key becomes a matter of trying to figure out, instead, how to boost the coverage of whatever you deem important against its prior performance."[18] Haile and the Pew Research Center, among others, have noted that engagement with news content on social media is not necessarily associated with deep reading or watching. In fact, access through social media is often correlated with lower time spent on a story as audiences graze and then click away quickly.[19]

For years, publishers and advertisers alike have condemned the cost per (thousand) mille (CPM) model and its derivatives (bulk measures of clicks or pageviews or, more recently, impressions), but little consensus has emerged around better alternatives. Still, there is some cause for optimism in this domain. In reviewing the research literature on the issue of analytics and metrics in journalism, Rodrigo Zamith has found that scholars who study news organizations' practices are seeing improvement and less purely clickbait-oriented work than was initially expected. Researchers are "observing more nuanced, if not restrained, attitudes, behaviors, and impacts on content, leading to presumptions of effects that are more limited than originally anticipated."[20]

Meanwhile, Facebook and Google, which can precisely target ads to relevant potential consumers, have moved in and dominated the online ad space. As outlets such as the *New York Times* and the *Washington Post* have begun to make substantial money from digital subscriptions, and news nonprofits such as the *Texas Tribune* have pioneered supporter and membership models, the conversation in media circles about how much raw traffic and "reach" are the ultimate goal has stalled a bit. Yet solving the mystery of going viral remains a top priority for digitally native publishers who know that building a large, loyal follower base is the key to any monetization strategy.

Seeing wide public circulation of content should also, under one theory of engagement, be something that serves journalism's democratic mission. As Matthew Hindman argues, "In order to maximize its social impact, ethical journalistic practice now requires strenuous attention to data on audience behavior." This new framework for "ethical metrics," he explains, means that a chief aim of journalism might become to "maximize the audience for civically valuable content."[21]

The larger point—that journalism has an ethical obligation to engage in networked practice or indeed to fuel networks themselves—is crucial. Society's news needs, as well as democracy's, are changing. Journalism must rise to meet them. By doing so, journalism will ensure its value and vital role in society long into the digital future.

9 News and Democracy: A Research Guide

This final chapter provides a granular, data-driven, and historically contextual orienting picture to help facilitate deeper background understanding. Any picture of our many overlapping media ecosystems together is necessarily complex, with areas of dramatic rupture relative to the past standing beside areas where change has only been incremental. The chapter is meant to stand as a useful guide to the social science research on many of the major points of interest and debate about changes in media and how they are affecting society. It is the backdrop against which journalists will attempt to apply knowledge and foster networks of recognition. As the data make clear, achieving these goals will continue to be no easy task, but we must understand these conditions for the modern news consumer and journalist, in all of their complexity.

Distrust in News

The General Social Survey (GSS) and Gallup have tracked public opinion about news media for long enough intervals to reveal definite, secular shifts in the American communications landscape. We have entered a new era in terms of how the public evaluates news media and its representations of facts, although these patterns need to be set in context. At the very moment professional, reportorial US media outlets have arguably become *more accurate* and more transparent than at any point in history, a substantial portion of the public seems to distrust and cast doubt on news institutions.[1] This may be because opinion journalism has increasingly been intermingled with, and conflated with, reportorial journalism—but there may be more going on than that.

In 1973, according to the General Social Survey, 23 percent of Americans had "a great deal" of confidence in the press, 62 percent had "only some," and 14 percent had "hardly any." These numbers held fairly steady through the immediate few years after Watergate, which saw the press play a heroic role in exposing corruption in the Nixon administration. Yet the negative trendline over the following four decades was nearly inexorable. By 2014, only 8 percent of those surveyed had "a great deal" of confidence in the press, while 46 percent had "only some," and, importantly, 44 percent said they had "hardly any," a figure that grew to 49 percent by 2016.

Likewise, Gallup data shows similar trends over the same multidecadal period: In 1972, a combined 68 percent of Americans had trust and confidence in mass media—a "great deal" (18 percent) or a "fair amount" (50 percent). By 2016, that aggregate figure had fallen by more than half, to 32 percent, with only 8 percent saying they had a great deal of trust and confidence. Negative evaluations of mass media had also increased more than fourfold, according to Gallup, from 6 percent in 1972 to 27 percent in 2016.

These forty years of American media history were vertiginous in their degree of change and innovation. Included were many milestones: the rise of cable news (CNN came to prominence in 1991; Fox News was founded in 1996); the end of the Fairness Doctrine in 1987; the rise of Internet browsers and digitally native news sites in the mid-1990s; and the founding of Facebook (2004) and Twitter (2006). These structural changes may generate additional semantic meanings for *news*, *media*, or *press* that complicate longitudinal survey patterns. Simply put, there are many more kinds of outlets—some purveying infotainment, others gossip and invective—that may get lumped, in the public's mind, into professional, reportorial journalism (see figure 9.1).

Are semantics then at fault here? As with GSS's slightly dated reference to the "press," Gallup's wording of questions emphasizes the traditional news media: "In general, how much trust and confidence do you have in the mass media—such as newspapers, T.V. and radio—when it comes to reporting the news fully, accurately, and fairly—a great deal, a fair amount, not very much, or none at all?" Yet the negative trend cannot be entirely explained away as an artifact of survey question wording and changes in the shared public meaning of *news*. In a broad analysis of nearly all journalism- and media-related survey data in recent decades, Lee B. Becker and Mengtian Chen conclude, "The public is very critical of the journalists and of the news media. It hardly matters what question is asked."[2]

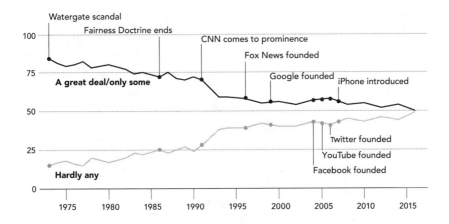

Figure 9.1
Media changes and confidence in the US press, 1973–2016.
Source: Survey data from the General Social Survey.

Still, how exactly Americans are updating their impressions of press institutions writ large based on the continuous churn and innovation in "new"
media is not exactly clear. Certainly, it is worth noting the dramatic decline
in daily consumption of newspapers over this period: the GSS found in 1972
that 68 percent of respondents read the newspaper each day, a number that
had fallen dramatically to 25 percent by 2014. It would be plausible that the
advent of partisan cable news shows and myriad news-oriented websites,
some overtly biased and others of varying quality, might be factored into
people's beliefs about the media. Yet such an interpretation is complicated
by the unrelenting downward trend, irrespective of technological change
over four decades. Indeed, the GSS data show that the number of Americans
who had "hardly any" trust in the press doubled between 1972 (14 percent)
and 1991 (29 percent), a period predating the ascendancy of cable news or
the Internet as mass communications phenomena. Although it is true that
conservatives express lower levels of trust in news media than do liberals,
overall declines in trust also are being driven by more negative evaluations
by both political independents and liberals.[3]

Another possible mitigating factor to consider is the degree to which
many other institutions in American society also have seen a parallel erosion
in public trust. Perhaps we are just in an antiauthority era. Yet other public
institutions such as the military, the judiciary, and the scientific community

have retained much greater trust relative to media. One piece of good news is that media distrust may, at least, be slowing down. Using a method to examine media patterns relative to the decline in trust in other institutions, Becker and Chen note that "media criticism is moderating somewhat relative to other institutions and occupations in the more recent years."[4]

Political communication scholars have sought to pinpoint the biggest drivers of media distrust. Among the many possible explanations that have been explored and to some extent eliminated are declining quality in media as a result of diminishing resources, a more openly hostile and adversarial media, a more negative news media, and a more biased media (and a public distaste for bias).[5] Two factors that appear to be most powerful, however, in generating negative opinions of news media are the rise of a tabloid style of coverage—defined by the political scientist Jonathan M. Ladd as "conventional news outlets' coverage of celebrities, sex scandals, and other topics once largely confined to tabloids"—as well as "elite criticism" from both Democratic and Republican leaders of news media institutions. In experimental work, at least, tabloid framings of news and criticisms by partisan leaders seem to be especially potent in generating negative feelings toward news media.

The obvious blurring of entertainment and news (linked to tabloidization) has also been lamented for a couple of decades now, and that trend certainly has a role in producing diminished credibility. One nuance in this regard should be noted: The trend may not be unequivocally negative for democracy. Political scientist and media scholar Matthew Baum has found that a more entertainment-driven, "soft news" style can serve to engage wider audiences than would normally pay attention to, for example, foreign policy issues—and soft news might even be a vehicle to inform.[6]

One of the most troubling recent patterns, accelerated in the era of President Trump, who has proven a fierce critic of news media, has been the growing partisan divide on the usefulness of the media's *watchdog*, or government accountability, role in American democracy. Of the many functions that the press plays, this is often considered its most important civic function and one the energies of which have been directed at Democrats and Republicans alike. However, the Pew Research Center, in a May 2017 survey, found that though 89 percent of Democrats think that "criticisms from news organizations keeps political leaders from doing what shouldn't be done," only 42 percent of Republicans believe this. Evaluations fluctuate as new administrations come into power, the report's authors acknowledge, but since the Pew Research Center began asking the question in 1985 the "distance between the parties has never approached the 47-point gap that exists today."[7]

"Small and Unelected Elite"

Where exactly did the disdain for mainstream news come from? A catch-all term that came into currency in the 1970s to capture both traditional broadcast and print together, *the media* was a kind of Rorschach blot drawn in some degree by the Nixon administration, which first began referring to "the media" while sharply criticizing its purported power over public opinion that allegedly surpassed that of any elected government.[8] Conjuring an image of conspiracy and hidden motive that endures, Vice President Spiro Agnew would famously say that media power was wrought by the "hands of a small and unelected elite."[9] President Trump's sustained rhetorical assault on these same institutions is of a piece, then, with a pattern begun in the 1970s, although he has inarguably taken it to new levels.

In the following decades, each successive presidential administration would in its own way attack and blame news media institutions—as a too-convenient scapegoat in some cases, and deservedly so in others—thus fueling mistrust in the public mind about news institutions. Deeper currents also helped compound a phenomenon that scholars have called the *hostile media effect*, whereby partisans of all sides would increasingly judge news institutions as unfavorable toward their values, beliefs, and policy preferences.[10] After Watergate and Vietnam, news media members began seeing themselves as much more openly adversarial to power, a phenomenon that was nominally healthy compared to the instincts of the more sycophantic press of the 1940s and 1950s, but which also resulted in a tendency to search for negative sound bites regardless of the truth of the matter. There became a "he said, she said" dimension to almost every issue, and no one in public life could anymore make claims unchecked by an equal and opposite negative reaction.[11]

Over those same decades, the academic left built a very deep and sophisticated set of critiques of media practice and its tendency to support power structures, fuel neoliberalism, and perpetuate hegemony and corporatist dominance. In a very different way, the political right has built its own body of critique. Although it lacks the coherence of the left's criticisms, it has more than made up for any deficiencies by establishing practical countermeasures—namely, an entire alternative media system, embodied by and anchored in Fox News, that rivals and even exceeds the likes of NBC and the *New York Times* in its scale of audience and deeply rooted persuasive power. We do not need to assign an equivalence here to state an obvious fact: both systems of media criticism, left and right, have made the job of the professional reporter simply harder to do, given the embedded skepticism of audiences.

Other long-term patterns were helping to consolidate certain overall impressions. The United States has steadily seen the "nationalization" of news media, as local and regional outlets, and news items, faded into the background and citizens were fed more news of general, Washington, DC-centric political and social import.[12] News reports also began moving away from event-centered or realist news—the basic who, what, when, where, and why—and toward "meaning-centered" news, with analysis and explanation increasingly common.[13] As mentioned, the presence of soft news and infotainment has increased in national outlets. This has coincided with tabloidization and an increasing tendency to focus on personalities and private lives in political and public affairs reporting.[14]

Telecommunications and technological shifts—from the rise of cable news and the advent of the World Wide Web in the 1990s, to the widespread adoption of social and user-generated media platforms such as Facebook, founded in 2004, YouTube (2005), and Twitter (2006)—also substantially expanded the possible meanings of *the media*, creating a kind of semantic gumbo. Many of these changes have fueled pockets of extremism on the ideological right; it is an ecosystem that in recent years has proved congenial to traditionally marginalized, more far-right voices and allowed them to play a more substantial role in shaping the conservative agenda.[15]

Media Fragmentation, Polarization

Although high levels of general distrust are being registered, there are relatively few people who would say they distrust their own preferred sources of news. It is worth recalling here the well-known paradox of Congressional retention (also called Fenno's paradox, after the political scientist Richard Fenno), whereby citizens generally disapprove of Congress but often approve of their local congressman/woman, leading to the puzzling situation whereby an institution with low approval ratings consistently sees its members reelected.[16] This paradox is at work with respect to news media.

The Pew Research Center has noted patterns in this regard. In a survey, consistent conservatives trusted Fox News at high levels (88 percent), while consistent liberals trusted NPR (72 percent), PBS (71 percent), the BBC (69 percent), and the *New York Times* (62 percent) at significant levels. Interestingly, however, respondents overall expressed more trust than distrust across the thirty-six news outlets included in the survey, and ABC News, NBC News, and CNN saw 50 percent or more of total respondents express trust in their work. Outlets such as the *Economist* and the *Wall Street Journal*, although not widely consumed necessarily by most Americans, saw much higher levels of trust than distrust among those who knew of the outlets.

Why do these levels of distrust matter, and what are their effects? Ladd has provided a succinct account: "Overall, media distrust leads to substantial information loss among the mass public," he notes. "Those who distrust the media both resist the information they receive from institutional news outlets and increasingly seek out partisan news sources that confirm their preexisting views. As a result, these individuals are less responsive to national policy outcomes, relying more on their political predispositions to form beliefs and preferences."[17] There is, in other words, a kind of negative feedback loop between the erosion of trust and the desire to seek out partisan news sources that reaffirm citizens' prior beliefs.

Political scientist James Campbell has argued that polarization should not be seen as something that has been "imposed" on the American populace, whether by parties, elites, or news media. Partisan media outlets, in this way, are capitalizing on an organic shift in attitudes and tastes that is antecedent to conditioning by news media itself; without shifts toward polarization deeply embedded in the fabric of democracy, such partisan outlets would lack an audience and die out.[18] Partisan media can fuel and sustain polarization, but the point is that such media may not be the primary or original cause.[19] Other research has emphasized that polarization is not just a phenomenon among elites, and it has been steadily rising across the country since at least 1972, accelerating in particular after 1992.[20] Careful analysis of the newspaper industry—seen as the bastion of traditional news values as compared to cable news—has shown that changes in consumer preferences likely exert strong influence on the partisan slant of news.[21]

Mass Third-Person Effects

Another factor unique to our era is relevant in this discussion of mistrust: the sheer volume of partisan content that circulates in the digital ecosystem, as well as an increased tendency by partisan media outlets to feature and critique messages from the opposing side—a kind of ubiquitous media criticism. Not only are citizens incidentally encountering messages from all sides more frequently through social and mobile channels, but also prominent broadcast hosts who have defined the beginning of the partisan broadcast era—take Bill O'Reilly, Sean Hannity, Rush Limbaugh, Jon Stewart, and Rachel Maddow, for example—as well as other types of partisan media have increasingly focused on "exposing" the tactics and messages of the other side. In this way, the strategy of much partisan media has become relatively less about advancing a policy agenda and more about surfacing and criticizing oppositional media messages. The net outcome is that the public has a great deal of visibility into the messages and news of different ideological camps, albeit a highly filtered view and interpretation.

Why does this increased visibility into other ideological worlds matter? More than three decades ago, in a landmark paper in communication and sociological research, W. Phillips Davison proposed what he called the "third-person effect hypothesis," which runs as follows:

In its broadest formulation, this hypothesis predicts that people will tend to overestimate the influence that mass communications have on the attitudes and behavior of others. More specifically, individuals who are members of an audience that is exposed to a persuasive communication (whether or not this communication is intended to be persuasive) will expect the communication to have a greater effect on others than on themselves. And whether or not these individuals are among the ostensible audience for the message, the impact that they expect this communication to have on others may lead them to take some action. Any effect that the communication achieves may thus be due not to the reaction of the ostensible audience but rather to the behavior of those who anticipate, or think they perceive, some reaction on the part of others.[22]

Davison formulated this hypothesis to account for a variety of phenomena that took place in an era of more limited media choice and access to information. The pervasive access and exposure many Americans now have to persuasive media from many other ideological communities and points of view may be fueling a sense of pessimism about others in society and

negative evaluations of the mediated public sphere. It may even be one factor that continues to drive, for example, the decline in trust of others, a pattern documented by the General Social Survey and other research surveys.[23]
As discussed previously, one of the most striking and unambiguous trends
in recent years is that partisans have become much more hostile in their
views toward the other side, and partisans perceive much more polarization
(even as the reality is more modest).[24]

The era of the social web offers unprecedented access to other points
of view, and though many people may engage in selective exposure to
conforming viewpoints, it is nearly impossible these days to avoid seeing
streams of media targeted to other ideological communities. The American
Press Institute found in 2014 that fully 60 percent of people say it is easier
to keep up with the news than five years ago; further, only about a quarter
of news consumers report paying for news. One reasonable interpretation
of these two survey findings is that news media are being pushed from
many directions and sources for average people, who may have little loyalty
to particular outlets because of a lack of economic commitment and attachment. Technology is fueling the complexity of what the American Press
Institute report calls the "personal news cycle," meaning that "the majority
of Americans across generations now combine a mix of sources and technologies to get their news each week."[25]

The third-person effect is being activated more frequently because of the
volume and diversity of digital media and social facts available. The consequence is a more regularly reinforced sense that highly susceptible "other"
persons in society are being persuaded of things that may be misguided or
repugnant. For the first time in human history, we are all afforded real-time
updates on, and windows into, the persuasive media of nearly all political
communities. The result is a collective sense of dread that others are being,
in effect, brainwashed or manipulated (even if "others" are actually exercising reason and are much less susceptible to persuasion than believed).

To return to Davison's overarching idea, one of the main effects of exposure to persuasive media intended for others is that the viewing subject
then takes action—for example, expresses public outrage or fear, or takes
political countermeasures—based on an assumption that the original message will negatively influence the other parties in question. The third-person effect, activated at mass scale, is driving feedback loops of distrust.

Selective Exposure and Its Consequences

What are the consequences of Americans choosing increasingly among partisan outlets? A body of political science research has accumulated that explores the mechanisms of what is called *selective media exposure*, or citizens choosing certain outlets that frequently conform to their beliefs. The chief findings are quite nuanced.

Natalie Jomini Stroud observes that, although partisan media may contribute to some detrimental outcomes in terms of diminishing the development of an informed public with shared concerns, selective exposure does have an upside: "Partisan media contribute to a more democratic system by providing an impetus for political participation. As a frequently employed benchmark, political participation is an important component of a properly functioning democracy." Overall, Stroud finds, partisan media can be invigorating and energizing. However, she warns, not everyone consumes partisan media, and this emerging media reality may "compound gaps in citizen participation."[26]

Many of the great liberal political thinkers over the past two centuries—John Stuart Mill, John Dewey, Jürgen Habermas—have theorized the benefits of citizens being challenged by different and competing perspectives and of cross-cutting exposure to diverse ideological viewpoints. It is in some ways an article of faith that a healthy democracy needs a robust deliberative component, a commons or public sphere that is rich and varied in its voices. However, careful examinations of these dynamics, such as that undertaken by Diana Mutz, have noted a kind of tragic potential flaw or paradox in operation. She notes that "theories of participatory democracy are in important ways inconsistent with theories of deliberative democracy. The best possible social environment for purposes of either one these two goals would naturally undermine the other."[27]

It is not necessary to choose between participation and deliberation in some absolute sense to acknowledge this basic tension. The findings of Stroud and Mutz, among others, complicate any kind of clear normative view of media polarization. Partisan media may spur more political participation, at least among partisans, while at the same time suppressing "big tent" deliberation across more ideologically diverse groups.

The direction of these findings is extended in the analysis of Kevin Arceneaux and Martin Johnson, whose experimental work on media choice and ideological broadcast channels leads them to conclude that the "rise of the partisan media on cable television is not the likely culprit" in polarizing Americans.[28] Persons who choose to consume partisan cable news content that conforms to their prior beliefs may be moved to more extreme positions—a phenomenon that scholars such as Cass Sunstein have lamented.[29] Yet the effects as measured so far are somewhat modest. Further, there is little evidence that forcing people to watch content that conflicts with their views has moderating effects. "If anything, it can be just as polarizing as exposure to proattitudinal [conforming] news," Arceneaux and Johnson note.[30]

The work of Markus Prior has generated further insights into the consequences of these media choices. American society, he argues, is returning in a way to a prenetwork television era, when fewer among the less educated segments of society were part of the news audience. The era of media choice means that only individuals with greater motivation to seek political news actually find and consume it. The rest seek entertainment of various kinds—and there is now ample supply across cable and the Internet—heightening what Prior terms "inequality" in political knowledge and involvement, and polarizing elections.[31]

Recent research on the 2016 election cycle indicates that a more polarized media ecosystem, particularly driven by the rise of powerful right-wing online news outlets, can have effects across the spectrum of news, shifting the overall agenda for mainstream outlets.[32] Whether this pattern accelerates in the years to come will be important.

Of course, younger Americans are those who are consuming more news via social media—and younger voters went overwhelmingly for Hillary Clinton in the 2016 election. Social channels cannot fully explain the rise of the Trump voter, although the suffusion of social facts in the media ecosystem almost certainly pushed mainstream coverage in Trump's direction.[33] Producers for the big television networks saw the popularity of Trump's social media presence and gave him ample airtime accordingly. Increases in polarization have taken place in much more substantial ways with older voters, who tend to watch a good deal of television, particularly cable, and who tend not to use social media.[34] Cable television, particularly Fox News, likely accounts for a substantial amount of the observed increase in polarization.[35] In sum, it may be that social media are leading to ideological extremes and cleavages, but it is far from a simple case or mechanism in terms of explaining or driving recent political events and trends.

Civic Knowledge and Media

At the root of worries about a degraded media environment are concerns that the public is awash in misinformation that is brainwashing them and that Americans are not properly fulfilling civic obligations as a result. Yet one thing is clear. The American public is not that much different now than it was in the immediate past in terms of basic political knowledge and attention to civic affairs. Indeed, it is striking to go back to older surveys on political knowledge, even reports from the 1990s. A 1994 survey-based report notes that the "American public has not absorbed the basic facts of many major news stories of recent months. … A new nationwide survey by the Times Mirror Center found the public largely uninformed about major national and international news stories."[36] In 1995, Pew, which grew out of the Time Mirror Center, summarized political-knowledge- and public-attention-related survey findings since 1989 as follows: "We have learned that relatively few serious news stories attract the attention of a majority of adult Americans, excepting those that deal with national calamities or the use of American military force. The average story tested by the Center was followed very closely by only one in four (25%) respondents."[37]

Further, in 2001, a broad review of the existing scholarly literature on political knowledge concluded: "Despite huge increases in the formal educational attainment of the U.S. population during the past 50 years, levels of political knowledge have barely budged. Today's college graduates know no more about politics than did high school graduates in 1950."[38] In 2015, Pew again conducted a knowledge test of a representative set of Americans and found substantial gaps in political knowledge, although education levels were highly correlated with knowledge. Nostalgia for a time when Americans were more uniformly informed—when Walter Cronkite told the nation, "That's the way it is," and the masses purportedly absorbed much more political knowledge—may therefore rest on false assumptions.[39]

There are nuances in this regard and slightly contrary views. Prior finds that "political knowledge has decreased for a substantial portion of the electorate: entertainment fans with access to new media." During earlier eras, a large subset of the population engaged in a kind of passive learning and acquisition of political knowledge because there was "simply nothing else to watch"; people consumed the news not out of strong interest, but because of a lack of choice. Our current situation, offering vast media choice, hollows out that incidental learning for a section of society. Still, Prior notes that knowledge levels in the aggregate have always been quite low, and the salient change in the cable and Internet era is a knowledge inequality between news junkies (who are found among all ideological groups) and entertainment fans.[40]

There is a kind of persistence in the American public's lack of knowledge about, and attention to, public affairs. This stasis is disturbed only by the occasional jolts of war and international crisis, with even a 9/11 terrorism attack type of event leading only to short-lived effects in terms of knowledge and attention.[41] This is not to suggest that the quality and power of the media industry makes no difference, but as with the issue of the public's trust in the press it appears there are structural factors at work that in some sense are deeper than, and antecedent to, changes in news media.

The evolving story of fake or false news surely will also continue to play a role in terms of knowledge acquisition by the public. What precisely are the kinds of fake news we've seen to date? Claire Wardle, who runs First Draft, a network dedicated to fighting misinformation and improving verification in media, has presented six types: authentic material used in the wrong context, imposter news sites designed to look like brands we already know, fake news sites, fake information, manipulated content, and parody content (that is misappropriated or misinterpreted).[42] Many observers have continued to call for greater action by the social media platforms in terms of filtering out such content. The success of such efforts can only be determined over the long run, as an "arms race" is sure to produce new tactics and attempts to spread false news.

In one of the earliest empirical studies of the phenomenon to follow the 2016 election, Hunt Allcott and Matthew Gentzkow conducted a number of experiments to test the effects of fake news. The paper concludes that the "average US adult might have seen perhaps one or several news stories in the months before the election."[43] More intriguingly, however, their survey experiments in the study seem to suggest that about 8 percent of persons surveyed will believe basically anything. Allcott and Gentzkow showed actually published fake news stories along with "placebos," or artificial fake news stories, to survey respondents: in each case, about 15 percent of people recalled seeing the story, actual or invented, and 8 percent believed it. The point here is that about 8 percent of the population, about one in twelve people, are susceptible to hoaxes and are willing to believe almost any outrageous claim if it conforms to their prior beliefs.

Some of the only noncomputational remedies that seem to have promise involve forewarning audiences of potential lies and distortions.[44] This might also include educating members of the public to spot sophistry and the tactics of disinformation. It is a strategy of inoculation, so to speak. In terms of educating the public about the persuasive and rhetorical techniques used to modify perceptions and shape public opinion, some researchers say it is worth harkening back to classic theories about propaganda, a field of study that has a rich history dating back to the Cold War, World War I, and World War II, and even further back in time.[45]

Overall, the technologies used to access news may be underappreciated in their contribution to risks in this area. The circulation of digital news through a blend of social media posts and feeds, texts and messaging applications, search engines and other algorithmically influenced applications creates the strong possibility for the weakening of public awareness of the exact sources of information. This opens the door for more misinformation to cycle through the media ecosystem. Indeed, as the Pew Research Center has noted, persons who consume digital news can remember the source of stories only about half of the time (56 percent).[46]

Losses in News and Civics

Compounding the changes that are inherent to an era of media choice, there is evidence that a contracting news media industry may have negative consequences for civic life at the local level. Research by Lee Shaker found that the closing of the *Rocky Mountain News* and *Seattle Post-Intelligencer* (which stopped issuing a print edition) newspapers was associated with diminished levels of civic engagement.[47] Another study also looked at the closing of a newspaper in Cincinnati and found fewer candidates running for office and lower voter turnout.[48] There has been much anxiety about what the FCC calls the "information needs of communities" and the consequences of a contracting news media.[49]

Indeed, Danny Hayes and Jennifer L. Lawless have studied local newspaper coverage patterns relating to congressional races and conclude that diminished local coverage is correlated with lower knowledge and engagement levels by citizens. Because local news outlets are contracting in so many areas, and so many voting districts have become uncompetitive (with either Republicans or Democrats dominating), another negative feedback loop may be in evidence across US political life: "When elections are less competitive and when districts are served [only] by large newspapers, media coverage of U.S. House campaigns is impoverished. Each factor contributes to less, and less substantive, coverage. This diminished news environment, then, depresses political engagement. Citizens in districts with less campaign coverage are less able to evaluate their incumbent and not as capable of making ideological judgments about the candidates vying for office. They are also less likely to vote in the House election. These effects occur for people regardless of their level of political attentiveness."[50] Some researchers have even begun to characterize certain communities that lack basic information resources as *news deserts*, with particular concerns for lower-income areas.[51]

The trend most relevant to those in the media industry over the past decade has been the inexorable decline in employment, cutting in half the overall number of reporters and editors in the newspaper business in a rather short period. The number of all employees (business, staff, and industrial functions included) in the newspaper publishing industry has declined even more dramatically (see figure 9.2). It is surely one of the largest relative declines over a short period of any industry in American history, including the various manufacturing industries, such as steel production, whose steep decline beginning in the 1970s and 1980s has hollowed out small towns and cities across the United States. Outside of a few national or global markets such as New York and Washington, DC, the news industry itself has rapidly become something of a rust belt.

Much of this collapse is explained by the loss of advertising dollars, which have substantially migrated to Internet companies. The relatively swift consolidation of digital media and communications around a few monster companies and platforms—the gravitational pull of which seems to attract an increasingly large share of available advertising dollars—is the result of powerful network effects distinctive of the digital age. Specifically here we are talking about the likes of Google, Facebook, Apple, Amazon, and Netflix. "The free and open web, architected for equal access, is instead dominated by a few large media companies who, in turn, are dominated by a few large technology platforms," Nieman Lab editor Josh Benton observes.[52]

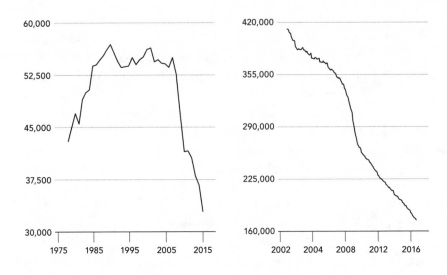

Figure 9.2
Decline in number of newspaper reporters and editors (left) and employment decline among all employees in newspaper publishing (right).
Source: American Society of News Editors (left); U.S. Bureau of Labor Statistics (right).

Among the most elite outlets, the name brands of which carry the promise of survival into the era of Internet-based, globalized media, there are occasional glimmers of promise, hints of staying-power. A few digitally native outlets, such as BuzzFeed, Vice News, and NowThis, have captured a share of the millennial audience and hope to keep them long into retirement; whether these new ventures will indeed be "hundred-year" companies is unknowable. Columbia University Journalism School Dean Steve Coll notes that a few big national newspapers and some of the new breed of nonprofit outlets are seeing renewal and recovery, but in cities and towns between the coasts there is an increasingly large "hole in the heart of American journalism."[53]

Conclusion: Fears and Solutions

This book has examined how the flow of news and information is changing in society and the role of social facts in shaping public knowledge and social epistemology. All the media-related changes I have described are happening in a political, emotional, and psychological climate that remains quite volatile. This volatility has something to do with the decentralized nature of our media ecosystem, which, as mentioned, is changing the structure of knowledge across many issues and domains. ICTs and decentralized digital networks fuel divergent civic epistemologies and lower the threshold for accessing them. This means that we can see, find, and be exposed to alternative, nonmainstream, and nonauthoritative sources of information with increasing ease.

With these shifts in mind, I have proposed a framework for news practice based around the ideas of knowledge-based journalism and generativity. I believe this framework could put news media in a better position to be relevant and succeed in the future. News practice should be centered on generating social connection. The idea of fostering networks of recognition can help clarify the role of media in this moment of transition. Journalists can help to bridge society's structural holes.

Our democracy needs more information pathways toward mutual recognition of shared issues, questions, struggles, and concerns. We need to be able to see and reflect on these mutual ties. Our media ecosystem sees plenty of viral content that is false, frivolous, or both. We need more civic virality.

Algorithmic Anxiety

This is no easy time to be a media consumer trying to make sense of the world. Citizens are grappling with a new world of media abundance and choice, accelerated by Internet-enabled mobile devices. We are quickly having to become a discerning set of consumers. Long gone is the age of limited choice, of three main broadcast networks—of Edward R. Murrow and Walter Cronkite, and the whole subsequent line of name-brand anchors, who centered the public's attention and knowledge. Now everyone must have a media opinion, must reflect on and justify her choices, even if her choice commonly is to disappear into the world of entertainment and avoid public affairs news as much as possible.

Much of the behavioral psychology and social science literature of recent decades tells us that humans are often poor decision makers when formulating choices under uncertainty. To boot, we have increasingly busy lives. Hardin's idea of an "economics of ordinary knowledge," as mentioned, operates in force.[1] We are very pragmatic about our news. Most of us have neither the time nor the energy to investigate news organizations and products thoroughly. And so we make the perfectly rational decision to either tune out or find a shortcut. Faced with a thousand outlets and a thousand apps, we turn to our social communities to help with these decisions. Our news habits are conditioned by social facts; and news itself, now washed around in a socially mediated environment of Facebook, Twitter, and the like, becomes a shared experience. This has consequences.

The novelist J. M. Coetzee once described how the mere awareness of a social setting changes the way one experiences the world. He was listening to the Goldberg Variations on the radio. He was generally enjoying the music, which he assumed was a studio recording, when he heard a coughing sound emitted from the broadcast. It was an audience member. Suddenly, Coetzee recalled, his experience of the music was transformed into rapture: "I could not see these fellow listeners, had no idea where they lived, who they were as individuals, but there was something obvious uniting us. ... For the duration of the performance we were, so to speak, one soul, united in—I can't find a better word—love."[2] There is something profound about the consumption of culture and information—whether music in a concert hall, a movie in a theater, or stories on a social platform—in the actual or implied presence of other human beings. The social experience transforms the moment from individual act to collective ritual. It raises the emotional stakes—both positive, as in the case of Coetzee's Bach experience, and negative (outrage or scorn), as often is the case with forms of news media.

It is interesting to consider that the trend of the increasing influence of social cues on public knowledge and cultural consumption predates the web. In *The Lonely Crowd* (1961), one of the twentieth century's most influential works of social commentary, author David Riesman observed that younger people were increasingly reading, listening to, and consuming stories and culture together in peer groups. This phenomenon, he posited, was influencing the American character so as to make people more *other-directed*, susceptible to social influence in their habits and values, as opposed to an orientation toward traditional values. This was increasingly true, Riesman said, even when people were alone: "One is almost always conscious of the brooding omnipresence of the peer-group."[3]

Nearly sixty years beyond *The Lonely Crowd*, we can see similar patterns at work, played out in an entirely new technological environment. Media is being more frequently injected with a kind of social super-infusion of emotion. We are in an age of ubiquitous social facts. More news items are coming to the attention of citizens through social channels. Information received through social ties frequently increases the salience of that information and potentially magnifies its emotional effect, as the phenomenon of social influence, particularly among strong ties, can be powerful. The net result is the increasing influence and power of news stories across the population.

The rise of social facts and the increasingly social experience of news engagement may account for a pervasive, albeit still anecdotal, feeling expressed by many persons in recent years that the world of media events seems to be spinning out of control, creating a sense of siege or fatigue.[4] Social-emotional fuel is being added to news through sharing behaviors, and the collective experience of news is more psychologically impactful and palpable for citizens.

The notion of fear is powerful in this regard: fear of others, fear of the messages they are hearing. There is a growing sense of the information sphere itself becoming highly contested, a battlefield or gladiatorial arena. Within this space, news and information are weaponized. Public awareness of media bias and fake news, and the technologies that allow for online manipulation, are heightening fear, feeding into the very notions of mistrust and distrust. Public fear and confusion then compound the problem.[5]

History can be some guide. Consider the dawn of the nuclear age. Anxiety manufactured its own social and psychological reality. In his 1950 Nobel Prize banquet speech, William Faulkner spelled out this reality in striking terms. The sense of fear—the brutally practical question, "When will I be blown up?"—had led to a retraction of common humanity and to a failure to grapple with deeper questions of the spirit. Addressing the "young man or woman writing today," Faulkner said the writer "must teach himself that the basest of all things is to be afraid; and, teaching himself that, forget it forever, leaving no room in his workshop for anything but the old verities and truths of the heart, the old universal truths lacking which any story is ephemeral and doomed—love and honor and pity and pride and compassion and sacrifice."[6] The character of fear and anxiety in our own era is different. It is too early to know precisely what this general fear and confusion is doing to us at some deeper level of psychology and the spirit. Yet undoubtedly, as Faulkner suggested, fear causes us to lose some access to a common humanity.

If the twentieth century were defined by a formula, it might have been $E = mc2$. It set in motion the nuclear age, the age in which humans held in their power, for the first time, their own destruction. The ensuing technology of that period provided a new way of understanding the universe and altered humans' view of their place in it. What formula is defining our age? The twenty-first century, the age of networks and data, has begun to be dominated by network analysis algorithms and graph analytics that help determine the importance of information and persons in networks. Depending on the domain or application, these may be metrics such as eigenvector centrality (which, as discussed, Google's engineers originally leveraged to create their search engine) or related concepts in network and graph theory.[7] Algorithms leverage such network measures to sort through, recommend parts of, and make selections in the sea of big data. Embedded in the computational processes of Google, Facebook, Amazon, and more, such network-analysis measures are being used to sort data across the Internet economy and the online social world. These formulas and the online platforms that ride on algorithmic pattern-finding are likewise beginning to alter our understanding of the world in a deeper way. They are affecting everything from news and commerce to dating and transportation.

Between Fragmentation and Consolidation

We may be in the midst of a new phase in a long-term pattern of American media cycles. We seem to be returning to a version of the prebroadcast world wherein decentralization prevails and citizen engagement with public affairs news is more unevenly distributed across the population. The period of mass media stability, when several television channels and established newspapers captivated the attention of the country's majority—a period running roughly from the immediate postwar years to the rise of cable and the Internet in the 1990s—is in this way a historical anomaly, characterizing but a small portion of US history.

There have been abiding tensions between impartiality and political bias in media, between degrees of centralization and distributed content production power, and between industrial media hierarchies and grassroots networks. "As the technology has changed," social scientist Gentzkow notes, "we're moving back to something that looks more like newspapers both in 1920 and in 1890, where there was much more diversity, much more competition and a much wider array of viewpoints catering to what people want."[8]

Many eras have seen media crosscurrents and hybrid regimes in these regards. For example, it is long forgotten now, but the Associated Press exercised monopoly control over the dissemination of news for many decades until 1945, when the Supreme Court intervened to stop the organization from mandating that its members exclude news organizations from any content sharing.[9] This monopoly operated prior to the big broadcast era; it was there even as more democratic revolutions broke out in the form of early, fragmented broadcasting and the information possibilities created by mass adoption of telephony. Crosscurrents have always been with us. Large corporate media mergers were taking place even as the early Internet was racing toward the democratization of publishing. Fears over a right-wing takeover of local television news in 2018 by a single corporate entity, Sinclair Broadcast Group, came even as commentators expressed worries about further fragmentation driven by social and online media.

Sociologist Paul Starr persuasively documents how significant public policy changes and commitments have been a driving force in generating a mosaic of societal and technological shifts, which in turn give birth to new forms of news and associated business models and market logics.[10] These constitutive choices spanned from the postal subsidies to early newspapers to regulatory shifts and spending priorities that aided the rise of the telegraph; policy choices profoundly affected radio, television, and the Internet in succession (while also reshaping their directions and modifying their public missions). Policy helped set the table for media innovation.

Since the Telecommunications Act of 1996, which provided the legal basis for the peer-to-peer platforms that have so fundamentally disrupted media, very little legislation of historic consequence has been passed in the areas of the Internet, media, and communications. The palpable sense of deep trouble with respect to media now, some two decades beyond the 1996 act, may result from a sense of policy drift—a sense of entropy that is ultimately destined to be reconciled by public policy changes in telecommunications that remain as yet unformulated. In our libertarian and antigovernment era, passing major new policies may seem inconceivable, but American history is replete with such policy interventions, which set in motion often unintended shifts in media. Concerns over data privacy may yet fuel new major legislation.

Internet and legal scholar Tim Wu has shown that American media and information industry empires have an almost regular pattern to their rise and fall: from early challenger to dominant monopoly to, inevitably, wounded incumbent disrupted by yet another new media challenger. When historians look back on our era, it is entirely possible that our moment will be seen as the end of a particular chapter characterized by digital disruption and fragmentation but ultimately followed again by another chapter of media consolidation.[11]

What exactly is the optimal balance for a democratic society between media fragmentation and consolidation? Ladd offers this formulation:

The United States should strive for a balance between a highly trusted, homogenous media establishment with little viable competition and an extremely fragmented media environment without any widely trusted information sources. In this middle path, individuals would continue to have a wide range of choices, including partisan, sensational, or conventionally objective news, as well as the option of avoiding news altogether. Yet the remaining institutional journalists and news outlets would continue to transmit important political information, with a significant portion of the public retaining enough confidence in the institutional press to use this information to hold government accountable.[12]

What should be emphasized as we potentially exit this era of media policy drift and fragmentation is that any potential interventions, public or private, have unknowable outcomes. Cable was once seen as a potentially democratizing and salutary phenomenon for education and democratic health; the Internet held the same early promise. Each has become in its way associated with degradation of deliberation and reason in the public sphere, whether or not such claims are fair (and I do not think the evidence entirely supports those claims). Any number of new policies could hold the seed of a new media revolution: regulation of social media companies, new wireless innovations, policies to unbundle cable packages, or some other yet unknown innovation, perhaps emanating from the realms of artificial intelligence research or virtual reality.

Although there are growing numbers of critics of the web and its effects on politics, many still hold out hope that a free and open Internet can produce a more informed citizenry capable of rational deliberation. How long this current media epoch will last remains to be seen. It is worth remembering that, if the past is any guide, it may take many, many decades for this current cycle to conclude and a different media paradigm to begin anew.

Supporting Networked Journalism

The discussion in this book has mostly avoided one of the biggest questions facing the media industry—namely, how to pay for news. I do maintain that more value-added, knowledge-driven news products and strategies will help media institutions differentiate themselves in the marketplace. Journalism must get away from redundant, commodity news.

From a financial perspective, the age of networks has meant one big thing for journalism: the drying up of advertising (and the failure of digital advertising to replace what was lost). This is because of the fundamental laws of networks and the "rich get richer" phenomenon that we discussed previously. It is also because data is the new currency of our age, and media companies lagged behind the likes of Google and Facebook in vacuuming up web user data. Journalism must now ponder a world in which advertising dollars are radically limited.

Although cycles of industry contraction and bankruptcy have always been seen in the news business, the United States enjoyed what looks in retrospect like a long period of relative growth, qualitative improvement, and stability in journalism. By happy accident, a societal good, public-interest journalism, was supported by an advertising model that made many media owners and journalists quite happy and reasonably well remunerated. That period, running from post-World War II until the early 2000s, is ending. It is therefore time to begin to think about a next stage in the development of news.

Media studies scholar and historian Victor Pickard notes that all too often, "future of news" discourse avoids any discussion of the long history of journalism policy interventions and shaping architectures, both through regulation and intentional laissez-faire inaction. He advocates that journalism studies as a "growing subfield should focus more on normative, policy-oriented, and economic questions pertaining to the future of news media institutions." There is ample historical evidence that such a focus on policy is necessary.[13] Starr, in his magisterial book *The Creation of the Media*, shows how, time and again, constitutive choices in the domains of technology policy, regulation, and law helped to set the predicate for the development of news media in the West.[14]

The American founders and their immediate predecessors sought to promote the diffusion of knowledge through the population by endorsing the postal subsidy, allowing newspapers and editorial content to circulate at much-reduced cost. "Whatever facilitates a general intercourse of sentiments," wrote James Madison in 1791, "as good roads, domestic commerce, a free press, and particularly a *circulation of newspapers through the entire body of the people* ... is equivalent to a contraction of territorial limits, and is favorable to liberty."[15] Madison made this observation as a way of arguing for postal subsidies to support the circulation of news periodicals.

During a subsequent chapter in American media's development, the advent of broadcast, the government sought to create a more regulated marketplace. But it took two decades to stand up a regulatory body that could sort out issues of spectrum allocation. Return again to the fallout, previously discussed, of the *Titanic* disaster in 1912: the Radio Act, and eventually the creation of the FCC in 1934. The FCC subsequently took on the role of enforcing some public interest obligations on the part of broadcasters and such rules as the now-defunct Fairness Doctrine and the equal-time provision.

For journalism to thrive in an age of networks and provide the kind of knowledge that society needs, there must be clear economic supports. It is a much better idea to provide subsidies, as with the postal discounts, than to try to get into the business of regulating the marketplace of ideas and news. A decentralized information ecosystem organized around the Internet is unlikely to countenance any second coming of some massive new regulatory regime.

In the wake of the 2016 election, there has been intense discussion of fake news and misinformation. Much of this has focused on the responsibility of Internet platform companies such as Google and Facebook to deprioritize sites that promote hoaxes, disinformation, or even propaganda. At root, one must admit, this is essentially a public pressure campaign to implore billionaire owners and their shareholders to voluntarily act in the public interest, a notion that private companies have little experience or expertise in defining.[16] There have been, to put it mildly, remarkably few bold ideas proposed at the structural level of public policy.

It would be beyond folly (and plainly unconstitutional) to call for restrictions on the press or on free speech. And yet, as constitutional law professor Cass Sunstein notes, "under current law in the United States (and generally elsewhere), government is permitted to subsidize speech however it wishes."[17] Further, Sunstein notes the founding tradition of the promotion of quality speech: not restrictions, but undergirding supports that are vital for the free speech tradition to thrive.

We find ourselves at a moment when our communications ecosystem may no longer be serving the interests of democracy. This realization is dawning on the public at the exact moment other massive shifts that could exacerbate the situation may be unfolding, including ongoing discussions of changing the rules relating to the Internet—specifically, to change the policy of *net neutrality*, which ensures that companies that control data flows on the Internet cannot favor or discriminate against websites. As such monumental changes loom, it is a vital moment to begin a big public conversation about news, information, and the future of democracy. Changes in net neutrality could substantially hurt news outlets of many kinds over the long run.[18] The cost of uploading and downloading data will have profound implications for the kind of networked journalism—the fostering of networks of recognition—that this book has advocated.

We are also at a moment when it has become clear that the nonprofit news sector can play a vital role in democracy. As part of a research project in 2017–2018 with my Northeastern University colleague Matt Nisbet, I interviewed more than thirty leading experts in this area, from foundation heads to the publishers of many news operations. About $1.8 billion in foundation funding poured into the nonprofit news sector (public media and universities included) between 2010 and 2015 in the United States.[19]

Much of this money flowed toward national-oriented outlets, PBS and NPR, and coastal cities, while the country's interior regions and local news start-ups there saw relatively little foundation funding. A tremendous amount of great national-oriented journalism was done as a result, but obviously the funding as a whole did little to change the overall political and cultural trajectory of the country, or substantially solve the deficits in local news, in those years leading up to the 2016 election.

To build back many of the institutions of journalism will likely take decades and a great deal more capital than is currently allocated. Community foundations and philanthropies will need to get much more involved, as a huge portion of existing support comes from a few dozen foundations. Even at the state and local level in particular, most nonprofit media funding derives from large national foundations, rather than community trusts or public charities and donor advised funds. Given that there is little tradition of giving to nonprofit media at these other types of grantmakers, they would benefit from partnering with experienced foundations.

In my conversations with thought leaders across the news nonprofit space, the problems of sustainability and funding remained front and center. New models must be built. There is general acknowledgment that nearly every state and region is now suffering from a deficit in news-generation capacity. As many of the examples detailed in this book have suggested, nonprofit news organizations are well positioned to do mission-driven work that is highly engaged with audiences. News nonprofits also can more readily define impact and engagement in such a way as to facilitate the network-building that represents the highest good journalism can provide.

A new framework is required for thinking about how a new system might operate to help support public-interest journalism. The goal must be to remove constraints on journalism that are currently imposed by the financial system—to make sure news is unencumbered both from a technological and a tax perspective as the advertising-supported model fails. Data is the fuel of the twenty-first century. How data is passed through the pipes of the Internet and at what cost will define success and failure for all forms of media in the foreseeable future. This has direct analogies to the postal system that was set up at the founding of the country to promote the diffusion of knowledge and news. The cost of delivery, its relative swiftness or slowness, is consequential for public knowledge.

If the policy of net neutrality is finally demolished, and data access becomes increasingly expensive, one possible solution that could help on many fronts would be to revive a kind of subsidy for news organizations within the broadband- and wireless-based telecommunications system. Conceptually, this might be framed around something we could call *news neutrality*, or *news-trality*. The idea would be for civic-oriented, public service news to be passed through communications networks with no associated cost. Obviously, there would be problems with determining who is eligible. Yet the courts, the tax system, the FCC, and the postal system all have experience in making distinctions in this domain. Moreover, making the system opt-in, not mandatory, would alleviate certain concerns. Policy details would need to be regularly renegotiated to account for the fast-moving technology sector, for which the future remains almost wholly unpredictable.

As Len Downie and Michael Schudson note in their important 2009 report and blueprint for the news future, "The Reconstruction of American Journalism," the government also needs to create much more clarity around how to achieve nonprofit and/or low-profit organizational statuses for news organizations and the rules of donations and revenue.[20] The current system is neither clear nor helpful. For years now, grassroots news startups have struggled to find a place under current tax rules, a situation that surely would have rankled those tax-resisting printers who launched the American Revolution and then enshrined related protections in the First Amendment.[21] It may be that a new kind of tax designation needs to be created to allow for-profit startup press organizations to thrive in a post-advertising era.

New ideas will inevitably need refinement and constant updating as technology races ahead, but we should not hesitate to begin creatively attacking the problems we see. The time is ripe. There is a palpable sense that we are at an inflection point with regard to news and its role in our democracy, as this book has suggested throughout. Even as distrust seems to prevail, there is a deep yearning for greater transparency, truth, and common facts. Tim Berners-Lee, the founder of the World Wide Web, has said: "There's a need for journalism. People are desperate for it. People are fed up with spam. They're fed up with just searching, using a web search tool to find a medical article, then realizing only after they have gone to the bottom of the article and followed the advice, and bought the drugs that the whole thing was produced by the same pharmaceutical company, with an extremely slanted view. People are getting so good at presenting stuff which is biased as though it is not."[22]

Changes at the structural level of media tend to take a long time to germinate. They make take decades or even generations. If the advent of broadcast, of radio and television, is any example, we are in for a long, complicated ride. The rise of the web and the digital revolution are again scrambling the rules of news and information. We find ourselves in what appears to be a time of chaos and factual recession. We also find ourselves in a moment that is a relative paradise for anyone seeking information about anything. The virtue of our information-communications ecosystem is deeply intertwined with its chief defect. Whether or not journalism is able to respond assertively to these challenges will determine how significant this defect eventually becomes—and if it might indeed, as some believe, threaten our democratic experiment.

Notes

Introduction

1. John Dewey, *The Public and Its Problems* (New York: Henry Holt and Company, 1927; reprint, Athens, OH: Swallow Press, 1954), 142.
2. Gallup/Knight Foundation, "American Views: Trust, Media and Democracy," 2017 Gallup/Knight Foundation Survey on Trust, Media and Democracy, January 16, 2018, https://knightfoundation.org/reports/american-views-trust-media-and-democracy.
3. For example, see Mike Ananny, "Networked Press Freedom and Social Media: Tracing Historical and Contemporary Forces in Press-Public Relations," *Journal of Computer-Mediated Communication* 19, no. 4 (July 2014): 938–956, https://doi.org/10.1111/jcc4.12076; Avery E. Holton, Seth C. Lewis, and Mark Coddington, "Interacting with Audiences," *Journalism Studies* 17, no. 7 (April 2016): 849–859, https://doi.org/10.1080/1461670X.2016.1165139; and Bregtje Van der Haak, Michael Parks, and Manuel Castells, "The Future of Journalism: Networked Journalism," *International Journal of Communication* 6 (2012): 1–20, http://ijoc.org/index.php/ijoc/article/view/1750/832.
4. David Streitfeld, "'The Internet Is Broken': @ev Is Trying to Salvage It," *New York Times*, May 20, 2017, https://www.nytimes.com/2017/05/20/technology/evan-williams-medium-twitter-internet.html.
5. James Madison, "Volume 5, Amendment I (Speech and Press), Document 24," *Founders' Constitution*, January 1800, http://press-pubs.uchicago.edu/founders/documents/amendI_speechs24.html.
6. Author interview with Matthew Gentzkow, November 30, 2017. Also see Hunt Allcott and Matthew Gentzkow, "Social Media and Fake News in the 2016 Election," *Journal of Economic Perspectives* 31, no. 2 (Spring 2017): 211–236, https://doi.org/10.1257/jep.31.2.211; Levi Boxell, Matthew Gentzkow, and Jesse M. Shapiro, "Is the Internet Causing Political Polarization? Evidence from Demographics," NBER Working Paper no. 23258, March 2017, https://www.brown.edu/Research/Shapiro/pdfs/age-polars.pdf; and Matthew Gentzkow and Jesse M. Shapiro, "What Drives Media Slant? Evidence from U.S. Daily Newspapers," *Econometrica* 78, no. 1 (January 2010): 35–71, https://doi.org/10.3982/ECTA7195.

7. Frederick F. Schmitt, ed., *Socializing Epistemology: The Social Dimensions of Knowledge* (London: Rowman & Littlefield, 1994).

8. Russell Hardin, *How Do You Know? The Economics of Ordinary Knowledge* (Princeton, NJ: Princeton University Press, 2009), 1–27.

9. Ibid., 4.

10. I use "social fact" throughout this book in various ways to describe patterns of online phenomena and signals in digital systems, as well as emerging trends in the nature of news production and how the public encounters and assimilates information. My usage here does echo Durkheim's original 1895 definition, in which he sought to classify a wide spectrum of phenomena for sociological study, from structural patterns of human life, norms, and laws to "social currents," "movements of opinion," and "transitory outbreaks" of human expression. See Émile Durkheim, *The Rules of Sociological Method*, trans. W. D. Halls (New York: The Free Press, 1982), 6.

11. Lilliana Mason, *Uncivil Agreement: How Politics Became Our Identity* (Chicago: University of Chicago Press, 2018).

12. Thomas E. Patterson, *Informing the News: The Need for Knowledge-Based Journalism* (New York: Vintage Books, 2013), 65–79.

13. Ibid., 7–8.

14. I have borrowed this term from law professor and technologist Jonathan L. Zittrain, who has used *generative* to describe open Internet and technology architectures that allow for creative and nonrestrictive use by the public. He defines this concept as follows: "Generativity denotes a technology's overall capacity to produce unprompted change driven by large, varied, and uncoordinated audiences." See Jonathan L. Zittrain, "The Generative Internet," *Harvard Law Review* 119, no. 7 (2006): 1974–2040, https://doi.org/10.1145/1435417.1435426.

15. John Wihbey, "Research Chat: Nicholas Lemann on Journalism, Scholarship, and More Informed Reporting," Journalist's Resource, Shorenstein Center on Media, Politics and Public Policy, updated December 7, 2011, https://journalistsresource .org/tip-sheets/research/nicholas-lemann-journalism-scholarship-reporting.

16. Robert G. Picard, "Deficient Tutelage: Challenges of Contemporary Journalism Education," paper presented at Toward 2020: New Directions in Journalism Education, Ryerson Journalism Research Centre, Toronto, Canada, May 31, 2014, http:// ryersonjournalism.ca/2014/03/17/toward-2020-new-directions-in-journalism -education/.

17. Of course, journalism has entered popular culture through other well-known characters and stories, such as Superman and its newspaper, the *Daily Planet*; Spider-Man as a photojournalist at the *Daily Bugle*; the comedy film *Anchorman*; and TV shows such as *Murphy Brown* and *The Mary Tyler Moore Show*.

18. Yphtach Lelkes, "Mass Polarization: Manifestations and Measurements," *Public Opinion Quarterly* 80, no. S1 (January 1, 2016): 392–410, https://doi.org/10.1093/ poq/nfw005.

19. Tim Alberta, "The Deep Roots of Trump's War on the Press," *Politico*, April 26, 2018, https://www.politico.com/magazine/story/2018/04/26/the-deep-roots-trumps -war-on-the-press-218105.

20. For example, see Robert W. McChesney, *The Problem of the Media: U.S. Communication Politics in the 21st Century* (New York: Monthly Review Press, 2004).

21. Walter Lippmann, *Liberty and the News* (New York: Harcourt, Brace and Howe, 1920), 5, https://archive.org/details/libertynews00lippuoft.

22. Commission on Freedom of the Press, *A Free and Responsible Press* (Chicago: University of Chicago Press, 1947), https://archive.org/details/freeandresponsib029216 mbp.

23. United States Kerner Commission, *Report of the National Advisory Commission on Civil Disorders* (New York: Bantam, 1968).

24. Peter Marks, "Word for Word/Al Gore's Senior Thesis; Even in College, the Vice President Was No Boob about the Tube," *New York Times*, July 16, 2000, https://www.nytimes.com/2000/07/16/weekinreview/word-for-word-al-gore-s-senior-thesis-even-college-vice-president-was-no-boob.html.

25. Advisory Committee on Public Interest Obligations of Digital Television Broadcasters, *Charting the Digital Broadcasting Future: Final Report of the Advisory Committee on Public Interest Obligations of Digital Television Broadcasters* (Washington, DC: The Benton Foundation, 1998), http://govinfo.library.unt.edu/piac/piacreport.pdf.

26. The Knight Commission on the Information Needs of Communities in a Democracy, *Informing Communities: Sustaining Democracy in the Digital Age* (Washington, DC: The Aspen Institute, 2009), https://assets.aspeninstitute.org/content/uploads/files/content/docs/pubs/Informing_Communities_Sustaining_Democracy_in_the_Digital_Age.pdf.

27. Steven Waldman and the Working Group on Information Needs of Communities Federal Communications Commission, *The Information Needs of Communities: The Changing Media Landscape in a Broadband Age* (Durham, NC: Carolina Academic Press, 2011).

28. David Foster Wallace, "Kenyon Commencement Address," lecture, May 21, 2005, https://web.ics.purdue.edu/~drkelly/DFWKenyonAddress2005.pdf.

29. Daniel Bell, *The Coming of Post-Industrial Society: A Venture in Social Forecasting* (New York: Basic Books, 1973), 263.

1   Digital Networks and Democracy's Needs

1. This section is based on Alison J. Head, John Wihbey, P. Takis Metaxas, and Dan Cohen, "Examining Information Processing Practices among Young News Consumers: A National Study to Inform Journalists, Librarians, and Educators," Project Information Literacy, October 2018, http://www.projectinfolit.org/news_study.html.

2. Luciano Floridi, *The Fourth Revolution: How the Infosphere Is Reshaping Human Reality* (Oxford: Oxford University Press, 2014), 40–41.

3. Ibid., 98.

4. Michael Schudson, *The Power of News* (Cambridge, MA: Harvard University Press, 1982), 3.

5. Ibid., 33.

6. Ibid., 18.

7. Christopher B. Daly, *Covering America: A Narrative History of a Nation's Journalism* (Amherst: University of Massachusetts Press, 2012), x–xi.

8. Matthew Gentzkow, Edward L. Glaeser, and Claudia Goldin, "The Rise of the Fourth Estate: How Newspapers Became Informative and Why It Mattered," in *Corruption and Reform: Lessons from America's Economic History*, ed. Edward L. Glaeser and Claudia Goldin (Chicago: University of Chicago Press, 2006), 187–230, http://www.nber.org/chapters/c9984.pdf.

9. Michael Schudson, "Question Authority: A History of the News Interview in American Journalism, 1860s–1930s," *Media, Culture & Society* 16 (1994): 565–587.

10. David Halberstam, *The Powers That Be* (Chicago: First Illinois, 2000), 31.

11. Paul Starr, *The Creation of the Media: Political Origins of Modern Communications* (New York: Basic Books, 2004), 28.

12. Floridi, *The Fourth Revolution*, 165.

13. Jeffrey C. Alexander, "Journalism, Democratic Culture, and Creative Reconstruction," in *The Crisis of Journalism Reconsidered: Democratic Culture, Professional Codes, Digital Futures*, ed. Jeffrey C. Alexander, Elizabeth Butler Breese, and Maria Luengo (Cambridge: Cambridge University Press, 2016), 22–23.

14. Jack Nicas, "YouTube Tops 1 Billion Hours of Video a Day, on Pace to Eclipse TV," *Wall Street Journal*, February 27, 2017; and Jim Edwards, "Planet Selfie: We're Now Posting a Staggering 1.8 Billion Photos Every Day," *Business Insider*, May 28, 2014, http://www.businessinsider.com/were-now-posting-a-staggering-18-billion-photos-to-social-media-every-day-2014-5.

15. Kevin Allocca, *Videocracy: How YouTube Is Changing the World … with Double Rainbows, Singing Foxes, and Other Trends We Can't Stop Watching* (London: Bloomsbury, 2018), xiv.

16. Thomas E. Patterson, "News Coverage of the 2016 General Election: How the Press Failed the Voters," Shorenstein Center on Media, Politics and Public Policy, December 2016, https://shorensteincenter.org/news-coverage-2016-general-election/.

17. Wolfgang Donsbach, "Journalism as the New Knowledge Profession and Consequences for Journalism Education," *Journalism* 15, no. 6 (July 2, 2013): 661–677, https://doi.org/10.1177/1464884913491347.

18. Steven Sloman and Philip Fernbach, *The Knowledge Illusion: Why We Never Think Alone* (New York: Riverhead Books, 2017).

19. Hardin, *How Do You Know?*, 4.

20. Sheila Jasanoff, *Designs on Nature: Science and Democracy in Europe and the United States* (Princeton, NJ: Princeton University Press, 2005).

21. Ibid., 270.

22. Albert-László Barabási, *Bursts: The Hidden Patterns Behind Everything We Do, from Your E-mail to Bloody Crusades* (New York: Dutton, 2010).

23. Adrienne Russell, *Networked: A Contemporary History of News in Transition* (Cambridge: Polity, 2011), 97–98.

24. Charles Taylor, "The Politics of Recognition," in *Multiculturalism: Examining the Politics of Recognition*, ed. A. Gutmann (Princeton, NJ: Princeton University Press, 1992), 26.

25. The concept of recognition has an extensive and complex intellectual history, and the word itself has a long genealogy. After identifying twenty-three different meanings of the word (in French), the philosopher Paul Ricoeur finds that there are three central meanings: recognition as identification, recognizing oneself (self-awareness or self-knowledge), and mutual recognition. See Paul Ricoeur, *The Course of Recognition*, trans. David Pellauer (Cambridge, MA: Harvard University Press, 2005).

26. Danielle Allen, *Talking to Strangers: Anxieties of Citizenship since* Brown v. Board of Education (Chicago: University of Chicago Press, 2004), 46–48.

27. Herbert J. Gans, *Democracy and the News* (Oxford: Oxford University Press, 2003), 55–61.

28. Emily Badger and Quoctrung Bui, "In 83 Million Eviction Records, a Sweeping and Intimate New Look at Housing in America," *New York Times*, April 7, 2018, https://www.nytimes.com/interactive/2018/04/07/upshot/millions-of-eviction -records-a-sweeping-new-look-at-housing-in-america.html.

29. James W. Carey, "A Short History of Journalism for Journalists: A Proposal and Essay," *Harvard International Journal of Press/Politics* 12, no. 3 (January 1, 2007), https://doi.org/10.1177/1081180X06297603.

30. Dewey, *The Public and Its Problems*, 180–181.

31. Henry Farrell, "New Problems, New Publics? Dewey and New Media," *Policy and Internet* 6, no. 2 (June 2014): 176–191, https://doi.org/10.1002/1944-2866.POI363.

32. Sharad Goel et al., "The Structural Virality of Online Diffusion," *Management Science* 62, no. 1 (July 22, 2015), https://doi.org/10.1287/mnsc.2015.2158.

33. Gary King, Benjamin Schneer, and Ariel White, "How the News Media Activate Public Expression and Influence National Agendas," *Science* 358, no. 6364 (November 10, 2017): 776–780.

34. John Wihbey, "The Challenges of Democratizing News and Information: Examining Data on Social Media, Viral Patterns and Digital Influence," Shorenstein Center on Media, Politics and Public Policy Discussion Paper Series D-85 (June 6, 2014), https://shorensteincenter.org/d85-wihbey/.

35. James G. Webster, *The Marketplace of Attention: How Audiences Take Shape in a Digital Age* (Cambridge, MA: MIT Press, 2014), 20.

36. Ibid., 160.

37. Mike Ananny, *Networked Press Freedom: Creating Infrastructures for a Public Right to Hear* (Cambridge, MA: MIT Press, 2018), 261.

38. James S. Ettema, "Journalism as Reason-Giving: Deliberative Democracy, Institutional Accountability, and the News Media's Mission," *Political Communication* 24, no. 2, (May 2007): 143–160, https://doi.org/10.1080/10584600701312860.

39. Steven Levitsky and Daniel Ziblatt, *How Democracies Die: What History Reveals about Our Future* (New York: Crown, 2018), 9.

40. Ibid.

41. See, for example, Robert D. Putnam, *Bowling Alone: The Collapse and Revival of American Community* (New York: Simon & Schuster, 2000); and Marc J. Dunkelman, *The Vanishing Neighbor: The Transformation of American Community* (New York: W. W. Norton & Co., 2014).

42. Theda Skocpol, *Diminished Democracy: From Membership to Management in American Civic Life* (Norman: University of Oklahoma Press, 2003), 231.

43. Christopher H. Achen and Larry M. Bartels, *Democracy for Realists: Why Elections Do Not Produce Responsive Government* (Princeton, NJ: Princeton University Press, 2016).

44. Ibid., 325; emphasis in original.

45. Patterson and Seib, "Informing the Public," in *The Press*, ed. Geneva Overholser and Kathleen Hall Jamieson (Oxford: Oxford University Press, 2005), 198.

46. Daniel Kreiss, "The Media Are about Identity, Not Information," in *Trump and the Media*, ed. Pablo J. Boczkowski and Zizi Papacharissi (Cambridge, MA: MIT Press, 2018), 96.

47. Ibid., 99.

48. Alexis de Tocqueville, *Democracy in America*, ed. J. P. Mayer (New York: Harper Perennial 1966/1988), 518.

49. These survey data are used in part in Meg Heckman and John Wihbey, "Mobile Site Experience and Local News: Missed Innovation Opportunities?," paper presented at the Association for Education in Journalism and Mass Communication (AEJMC), Media Management, Economics and Entrepreneurship Division, Norman, OK, March 2, 2018.

50. Alex S. Jones, *Losing the News: The Future of the News that Feeds Democracy* (Oxford: Oxford University Press, 2009).

2   News, Knowledge, and Civic Virality

1. My great thanks to Northeastern faculty colleagues Matt Carroll, Dan Zedek, and Walter Robinson, all of whom were at the *Globe* at the time of the investigation. Carroll and Robinson were key members of the Spotlight team that carried out the investigation. They helped with insights for formulating the graphic presented in figure 2.1. Zedek was instrumental in producing the visualization.

2. Members of the Spotlight team such as Matt Carroll and Walter Robinson recall email as a vital new factor in facilitating public engagement, creating feedback loops that contributed to their journalistic workflow and helping fuel national and global attention. Independent observers have also concluded that email was a decisive factor in the events. See, for example, Clay Shirky, *Here Comes Everybody: The Power of Organizing without Organizations* (New York: Penguin, 2008), 143–160.

3. Manuel Castells, *Communication Power* (Oxford: Oxford University Press, 2009), 19–20.

4. Matt Carroll, Sacha Pfeiffer, and Michael Rezendes, "Church Allowed Abuse by Priest for Years," *Boston Globe*, January 2, 2002, https://www.bostonglobe.com/news/special-reports/2002/01/06/church-allowed-abuse-priest-for-years/cSHfGkTIrAT25qKGvBuDNM/story.html.

5. J. David Goodman, "New York Police Twitter Strategy Has Unforeseen Consequences," *New York Times*, April 23, 2014, https://www.nytimes.com/2014/04/24/nyregion/new-york-police-twitter-strategy-has-unforeseen-consequences.html; and Jennifer Preston, "New York Police Twitter Backlash Spreads around the World," April 23, 2014, *New York Times*, https://thelede.blogs.nytimes.com/2014/04/23/new-york-police-twitter-backlash-spreads-around-the-world/.

6. Sarah J. Jackson and Brooke Foucault Welles, "Hijacking #myNYPD: Social Media Dissent and Networked Counterpublics," *Journal of Communication* 65, no. 6 (December 2015): 932–952, https://doi.org/10.1080/1369118X.2015.1106571.

7. Ibid. For more on Jackson's and Welles's ideas about counterpublics, see Sarah J. Jackson and Brooke Foucault Welles, "#Ferguson Is Everywhere: Initiators in Emerging Counterpublic Networks," *Information Communication and Society* 19, no. 3 (2016): 397–418, https://doi.org/10.1080/1369118X.2015.1106571.

8. J. David Goodman, "On Staten Island, Thousands Protest Police Tactics," *New York Times*, August 23, 2014, https://www.nytimes.com/2014/08/24/nyregion/on-staten-island-thousands-protest-police-tactics.html.

9. Bill Kovach and Tom Rosenstiel, *The Elements of Journalism: What Newspeople Should Know and the Public Should Expect* (New York: Three Rivers Press, 2014).

10. Tom Rosenstiel, "News as Collaborative Intelligence: Correcting the Myths about News in the Digital Age," Center for Effective Public Management at Brookings, June 30, 2015, https://www.brookings.edu/research/news-as-collaborative-intelligence-correcting-the-myths-about-news-in-the-digital-age/.

11. Marian F. MacDorman et al., "Recent Increases in the U.S. Maternal Mortality Rate: Disentangling Trends from Measurement Issues," *Obstetrics and Gynecology* 128, no. 3 (September 2016): 447–455, https://doi.org/10.1097/AOG.0000000000001556.

12. Nina Martin and Renee Montagne, "The Last Person You'd Expect to Die in Childbirth," ProPublica, May 12, 2017, https://www.propublica.org/article/die-in-childbirth-maternal-death-rate-health-care-system.

13. Author interview with Adriana Gallardo, June 27, 2017; and Adriana Gallardo, "What We've Learned So Far about Maternal Mortality from You, Our Readers," ProPublica, May 18, 2017, https://www.propublica.org/article/what-weve-learned-so-far-about-maternal-mortality-from-you-our-readers.

14. ProPublica and the *Virginian-Pilot*, "Reliving Agent Orange," series, ProPublica, 2016, https://www.propublica.org/series/reliving-agent-orange.

15. Rachel Grozanick, "Pulitzer Prize Winner Dave Philipps on Using Social Media to Find Untold Stories," The GroundTruth Project, July 11, 2017, http://thegroundtruthproject.org/pulitzer-prize-winner-dave-philipps-on-using-social-media-to-find-untold-stories/.

16. Angelica Das, *Pathways to Engagement: Understanding How Newsrooms Are Working with Communities*, Public Square Program, Democracy Fund, May 2, 2017, https://www.democracyfund.org/publications/pathways-to-engagement-understanding-how-newsrooms-are-working-with-communi.

17. Robert R. McCormick Foundation, "Lessons from #10YearsUP and Resources for Audience Engagement," accessed June 19, 2018, https://www.newsu.org/audience-engagement-resources.

18. Jane Elizabeth, "After a Decade, It's Time to Reinvent Social Media in Newsrooms," American Press Institute, November 14, 2017, https://www.americanpressinstitute.org/publications/reports/strategy-studies/reinventing-social-media/.

19. Ross Reynolds, "KUOW Uses Speed-Dating Format to Help People Understand Each Other," *Current*, March 6, 2018, https://current.org/2018/03/kuow-uses-speed-dating-format-to-help-people-understand-each-other/.

20. Jeremy Hay and Eve Pearlman, "Can People Be Civil About Polarizing Topics? 'Dialogue Journalism' Could Serve as a Roadmap," Poynter, March 1, 2018, https://www.poynter.org/news/can-people-be-civil-about-polarizing-topics-dialogue-journalism-could-serve-roadmap.

21. Christine Schmidt, "What Strategies Work Best for Increasing Trust in Local Newsrooms? Trusting News Has Some Ideas," Nieman Lab, February 16, 2018, http://www.niemanlab.org/2018/02/what-strategies-work-best-for-increasing-trust-in-local-newsrooms-trusting-news-has-some-ideas/.

22. News Voices, presented by Free Press, https://www.freepress.net/news-voices.

23. Patterson, *Informing the News*, 22–23.

24. Allen, *Talking to Strangers*, xxii.

25. Rasmus Kleis Nielsen, "Digital News as Forms of Knowledge: A New Chapter in the Sociology of Knowledge," in *Remaking the News: Essays on the Future of Journalism Scholarship in the Digital Age*, ed. Pablo J. Boczkowki and C. W. Anderson (Cambridge, MA: MIT Press), 106.

26. William James, *The Principles of Psychology*, vol. 1 (New York: Henry Holt & Co., 1890), 259. See also Robert E. Park, "News as a Form of Knowledge: A Chapter in the Sociology of Knowledge," *American Journal of Sociology* 45, no. 5 (March 1940): 669–686, https://doi.org/10.1086/218445.

27. Hardin, *How Do You Know?*, 1–27.

28. Ibid., 2.

29. There are even deeper and more complicated philosophical links between these two definitions of knowledge, and notions of public and private knowledge. But my concern here is not the more technical aspects of epistemology. See Dewey, *The Public and Its Problems*; and Ludwig Wittgenstein, *Philosophical Investigations*, 3rd ed., trans. G. E. M. Anscombe (London: Pearson, 1973).

30. James Gleick, *The Information* (New York: Pantheon Books, 2011), 151.

31. David Weinberger, *Small Pieces Loosely Joined: A Unified Theory of the Web* (New York: Basic Books, 2002).

32. See Stanley Milgram, "The Small-World Problem," *Psychology Today* 1, no. 1 (May 1967): 61–67, https://www.cs.purdue.edu/homes/agebreme/Networks/papers/milgram67smallworld.pdf; and Mark S. Granovetter, "The Strength of Weak Ties," *American Journal of Sociology* 78, no. 6 (May 1973): 1360–1380, https://doi.org/10.1086/225469.

33. Ronald S. Burt, "Structural Holes and Good Ideas," *American Journal of Sociology* 110, no. 2 (September 2004): 349–399, https://doi.org/10.1086/421787.

34. Pablo J. Boczkowski and Eugenia Mitchelstein, "Scholarship on Online Journalism: Roads Traveled and Pathways Ahead," in *Remaking the News: Essays on the Future of Journalism Scholarship in the Digital Age*, ed. C. W. Anderson and Pablo J. Boczkowki (Cambridge, MA: MIT Press, 2017), 22.

35. Sue Robinson, "Trump, Journalists, and Social Networks of Trust," in *Trump and the Media*, ed. Pablo J. Boczkowki and Zizi Papacharissi (Cambridge, MA: MIT Press, 2018), 193.

36. Duncan Watts, *Everything Is Obvious: How Common Sense Fails Us* (New York: Crown Business, 2011).

37. Albert-László Barabási, *Network Science* (Cambridge: Cambridge University Press, 2016), 109–110; and Duncan Watts, *Six Degrees: The Science of a Connected Age* (New York: W. W. Norton & Co.., 2004), 239–241.

38. Barabási, *Bursts: The Hidden Patterns Behind Everything We Do, from Your E-mail to Bloody Crusades* (New York: Dutton, 2010).

39. Dewey, *The Public and Its Problems*, 142.

40. James Carey, "A Cultural Approach to Communication," in *Communication as Culture: Essays on Media and Society* (Boston: Unwin Hyman, 1989), 13–36, http://web.mit.edu/21l.432/www/readings/Carey_CulturalApproachCommunication.pdf.

41. Dewey, *The Public and Its Problems*, 184.

42. Limor Shifman, *Memes in Digital Culture* (Cambridge, MA: MIT Press, 2014), 60–61.

43. Derrick Z. Jackson, "Environmental Justice? Unjust Coverage of the Flint Water Crisis," Shorenstein Center on Media, Politics and Public Policy, July 2017, https://shorensteincenter.org/wp-content/uploads/2017/07/Flint-Water-Crisis-Derrick-Z-Jackson-1.pdf?x78124.

44. M. B. Pell and Joshua Schneyer, "The Thousands of U.S. Locales Where Lead Poisoning Is Worse than in Flint," *Reuters Investigates*, December 19, 2016, https://www.reuters.com/investigates/special-report/usa-lead-testing/.

45. Herbet Gans, "Multiperspectival News Revisited: Journalism and Representative Democracy," *Journalism* 12, no. 1 (January 2011): 3–13, https://doi.org/10.1177/1464884910385289.

46. Paul Lazarsfeld and Robert Merton, "Friendship as a Social Process: A Substantive and Methodological Analysis," in *Freedom and Control in Modern Society* (New York: Van Nostrand, 1954), 18–66; and Miller McPherson, Lynn Smith-Lovin, and James M. Cook, "Birds of a Feather: Homophily in Social Networks," *Annual Review of Sociology* 1 (2001): 415–444, https://doi.org/10.1146/annurev.soc.27.1.415.

47. M. E. J. Newman, "Modularity and Community Structure in Networks," *Proceedings of the National Academy of Sciences* 103, no. 23 (June 6, 2006): 8577–8582, https://doi.org/10.1073/pnas.0601602103.

48. Lauren Kirchner, "When Discrimination Is Baked into Algorithms," *Atlantic*, September 6, 2015, https://www.theatlantic.com/business/archive/2015/09/discrimination-algorithmsdisparate-impact/403969/.

49. Mark A. Pachucki and Ronald L. Breiger, "Cultural Holes: Beyond Relationality in Social Networks and Culture," *Annual Review of Sociology* 36 (2010): 205–224, https://doi.org/10.1146/annurev.soc.012809.102615.

50. Rick Edmonds, "What Does Mark Zuckerberg Really Think of the Value of News? Not All that Much," Poynter, November 7, 2017, https://www.poynter.org/news/what-does-mark-zuckerberg-really-think-value-news-not-all-much.

51. Eli Pariser, "When the Internet Thinks It Knows You," *New York Times*, May 22, 2011, https://www.nytimes.com/2011/05/23/opinion/23pariser.html.

52. Jay Rosen, *What Are Journalists For?* (New Haven, CT: Yale University Press, 1999).

53. Joyce Y. M. Nip, "Exploring the Second Phase of Public Journalism," *Journalism Studies* 7, no. 2 (February 2007): 212–236, https://doi.org/10.1080/14616700500533528; and Jack Rosenberry and Burton St. John III, "Introduction: Public Journalism Values in an Age of Media Fragmentation," in *Public Journalism 2.0: The Promise and Reality of a Citizen-Engaged Press*, ed. Jack Rosenberry and Burton St. John III (New York: Routledge, 2010), 1–8.

54. Rosenstiel, "News as Collaborative Intelligence."

55. Dewey, *The Public and Its Problems*, 139–142.

3   Social Facts and Contested Knowledge

1. Soroush Vosoughi, Deb Roy, and Sinan Aral, "The Spread of True and False News Online," *Science* 359, no. 6380 (March 2018): 1146–1151, https://doi.org/:10.1126/science.aap9559.

2. Stephanie McCrummen, Beth Reinhard, and Alice Crites, "Woman Says Roy Moore Initiated Sexual Encounter When She Was 14, He Was 32," *Washington Post*, November 9, 2017, https://www.washingtonpost.com/investigations/woman-says -roy-moore-initiated-sexual-encounter-when-she-was-14-he-was-32/2017/11/09/ 1f495878-c293-11e7-afe9-4f60b5a6c4a0_story.html.

3. Sophie Gilbert, "The Movement of #MeToo," *Atlantic*, October 16, 2017, https:// www.theatlantic.com/entertainment/archive/2017/10/the-movement-of-metoo/ 542979/.

4. Daniel Strauss, "Moore Lashes Out at *Washington Post*," Politico, November 11, 2017, https://www.politico.com/story/2017/11/11/roy-moore-washington-post-244805.

5. Paul Starr, *The Creation of the Media: Political Origins of Modern Communications* (New York: Basic Books, 2004), 218–219.

6. "President Moves to Stop Mob Rule of Wireless," *New York Herald*, August 17, 1912. See also Mark Raboy, *Marconi: The Man Who Networked the World* (Oxford: Oxford University Press, 2016), 358–359.

7. Eric Klinenberg, *Fighting for Air: The Battle to Control America's Media* (New York: Metropolitan Books, 2007), 18–19.

8. Lincoln Caplan, "Should Facebook and Twitter Be Regulated Under the First Amendment?," *Wired*, October 11, 2017, https://www.wired.com/story/should -facebook-and-twitter-be-regulated-under-the-first-amendment/; and Peter Baker and Celia Kang, "Trump Threatens NBC Over Nuclear Weapons Report," *New York Times*, October 11, 2017, https://www.nytimes.com/2017/10/11/us/politics/trump -nbc-fcc-broadcast-license.html.

9. Russell Hardin, "The Crippled Epistemology of Extremism," in *Political Extremism and Rationality*, ed. Albert Breton et al. (Cambridge: Cambridge University Press, 2002), 3–22, https://doi.org/10.1017/CBO9780511550478.002.

10. There are a number of interesting historical parallels in this regard. For more, see Niall Ferguson, *The Square and the Tower: Networks and Power, from the Freemasons to Facebook* (New York: Penguin Press, 2018).

11. George W. Bush, "The Spirit of Liberty: At Home, In the World," remarks given at the George W. Bush Institute, New York, October 19, 2017, https://www .bushcenter.org/exhibits-and-events/events/2017/10/p-spirit-of-liberty.

12. Barack Obama, "Remarks by President Obama and Chancellor Merkel of Germany in a Joint Press Conference," remarks given at the German Chancellory, Berlin, Germany, November 17, 2016, https://obamawhitehouse.archives.gov/the -press-office/2016/11/17/remarks-president-obama-and-chancellor-merkel-germany -joint-press.

13. The phrase originally comes from Neil Postman, *Amusing Ourselves to Death: Public Discourse in the Age of Show Business* (New York: Viking Penguin, 1985).

14. Sue Halpern, "How He Used Facebook to Win," *New York Review of Books*, June 8, 2017, http://www.nybooks.com/articles/2017/06/08/how-trump-used-facebook-to -win/.

15. For a sense of the important work being done in this domain, see David Karpf, *Analytic Activism: Digital Listening and the New Political Strategy* (Oxford: Oxford University Press, 2016); and Daniel Kreiss, *Prototype Politics: Technology-Intensive Campaigning and the Data of Democracy* (Oxford: Oxford University Press, 2016).

16. Data set can be accessed at https://repository.library.northeastern.edu/ collections/neu:m0417171j.

17. Thomas Patterson, "Doing Well and Doing Good," Shorenstein Center on Media, Politics and Public Policy, 2000, https://shorensteincenter.org/wp-content/ uploads/2012/03/soft_news_and_critical_journalism_2000.pdf.

18. That analysis can be found at https://github.com/aleszu/topicmodeling/blob/ master/README.md. My special thanks to my Northeastern University colleague Aleszu Bajak for facilitating this analysis.

19. Liam Corcoran, "Get NewsWhip's Three Years of Social Data Report," News-Whip, March 1, 2017, https://www.newswhip.com/2017/03/three-years-social-data-report/.

20. Amy Mitchell et al., "The Modern News Consumer," Pew Research Center, July 7, 2016, http://www.journalism.org/2016/07/07/trust-and-accuracy/.

21. Ahmed Al-Rawi, "Viral News on Social Media," *Digital Journalism*, October 2017, 1–17, https://doi.org/10.1080/21670811.2017.1387062.

22. Daniel M. Romero, Brendan Meeder, and Jon Kleinberg, "Differences in the Mechanics of Information Diffusion across Topics: Idioms, Political Hashtags, and Complex Contagion on Twitter," in *Proceedings of the 20th International Conference on World Wide Web* (New York: ACM, 2011), 695–704 https://doi.org/10.1145/1963405.1963503.

23. Adam Mordecai, "9 out of 10 Americans are Completely Wrong About This Mindblowing Fact," Upworthy, March 4, 2013, http://www.upworthy.com/9-out-of-10-americans-are-completely-wrong-about-this-mind-blowing-fact-2.

24. Pablo J. Boczkowski and Eugenia Mitchelstein, *The News Gap: When the Information Preferences of the Media and the Public Diverge* (Cambridge, MA: MIT Press, 2015).

25. Joseph E. Uscinski, *The People's News: Media, Politics, and the Demands of Capitalism* (New York: New York University Press, 2014).

26. For example, see Felippe Rodrigues, "How the Associated Press is Experimenting with Headlines and Modular Stories to Win Facebook," Storybench, August 26, 2017, http://www.storybench.org/how-the-associated-press-is-winning-facebook/.

27. Elisa Shearer and Jeffery Gottfried, "News Use across Social Media Platforms 2017," Pew Research Center, September 7, 2017, http://www.journalism.org/2017/09/07/news-use-across-social-media-platforms-2017/.

28. Robert E. Park, Ernest W. Burgess, and Roderick D. McKenize, *The City* (Chicago: University of Chicago Press, 1967), 39, http://www.esperdy.net/wp-content/uploads/2009/09/Park-The-City.pdf.

29. Jon Kleinberg, "Information Flow and Graph Structure in Online Social Networks," lecture, Becker Friedman Institute at the University of Chicago September 24, 2016, YouTube video, https://youtu.be/iccgU1ul13E.

30. Patterson and Seib, "Informing the Public," 192.

31. David Weinberger, *Too Big to Know: Rethinking Knowledge Now that Facts Aren't the Facts, Experts Are Everywhere, and the Smartest Person in the Room Is the Room* (New York: Basic Books, 2012), 1–13.

32. Ibid.

33. Andrew Chadwick, *The Hybrid Media System: Politics and Power* (Oxford: Oxford University Press, 2013); and Webster, *The Marketplace of Attention*.

34. David Grewal, *Network Power: The Social Dynamics of Globalization* (New Haven, CT: Yale University Press, 2008).

35. Jasanoff, *Designs on Nature*, 270.

36. Nicco Mele, *The End of Big: How the Digital Revolution Makes David the New Goliath* (New York: St. Martin's Press, 2013).

37. Ben Smith and Byron Tau, "Birtherism: Where It All Began," *Politico*, April 22, 2011, https://www.politico.com/story/2011/04/birtherism-where-it-all-began-053563.

38. Josh Clinton and Carrie Roush, "Poll: Persistent Partisan Divide over 'Birther' Question," *NBC News*, August 10, 2016, https://www.nbcnews.com/politics/2016 -election/poll-persistent-partisan-divide-over-birther-question-n627446.

39. Charles McKay, *Memoirs of Extraordinary Popular Delusions and the Madness of Crowds* (London: Richard Bentley, 1841), https://www.cmi-gold-silver.com/pdf/ mackaych2451824518-8.pdf.

40. Geoffrey D. Munro et al., "Biased Assimilation of Sociopolitical Arguments: Evaluating the 1996 U.S. Presidential Debate," *Basic and Applied Social Psychology* 24, no. 1 (2002): 15–26, https://www.tandfonline.com/doi/abs/10.1207/S15324834B ASP2401_2; and Dan M. Kahan, "Ideology, Motivated Reasoning, and Cognitive Reflection," *Judgement and Decision Making* 8, no. 4 (July 2013): 407–424, http:// journal.sjdm.org/13/13313/jdm13313.pdf.

41. See the canonical study in this area: Solomon E. Asch, "Studies of Independence and Conformity: A Minority of One against a Unanimous Majority," *Psychological Monographs: General and Applied* 70, no. 9 (1956): 1–70, http://psycnet.apa.org/ record/2011-16966-001.

42. Joseph Marks et al., "Epistemic Spillovers: Learning Others' Political Views Reduces the Ability to Assess and Use Their Expertise in Nonpolitical Domains," Harvard Public Law Working Paper no. 18–22, April 13, 2018, https://ssrn.com/ abstract=3162009.

43. Caitlin Drummond and Baruch Fischhoff, "Individuals with Greater Science Literacy and Education Have More Polarized Beliefs on Controversial Science Topics," *Proceedings of the National Academy of Sciences of the United States of America* 114, no. 36 (2017): 9587–9592, https://doi.org/10.1073/pnas.1704882114.

44. Cass R. Sunstein and Adrian Vermeule, "Conspiracy Theories," John M. Olin Program in Law and Economics Working Paper no. 387, University of Chicago, Chicago, 2008.

45. Hardin, *How Do You Know?*, 185.

46. Ibid., 203–204.

47. Timur Kuran, "Availability Cascades and Risk Regulation," University of Chicago Public Law & Legal Theory Working Paper, no. 181, 2007.

48. Kate Starbird, "Examining the Alternative Media Ecosystem through the Production of Alternative Narratives of Mass Shooting Events on Twitter," Association for the Advancement of Artificial Intelligence, 2007, http://faculty.washington.edu/ kstarbi/Alt_Narratives_ICWSM17-CameraReady.pdf.

49. TwitterTrails, "Claim: Melania Trump Had an Exorcist in the White House," TwitterTrails, February 22, 2018, http://twittertrails.wellesley.edu/~trails/stories/ investigate.php?id=6911497280.

50. Brendan Nyhan and Jason Reifler. "When Corrections Fail: The Persistence of Political Misperceptions," *Political Behavior* 32, no. 2 (June 2010): 303–330, https:// doi.org/10.1007/s11109-010-9112-2.

51. Lucas Graves, *Deciding What's True: The Rise of Political Fact-Checking in American Journalism* (New York: Columbia University Press, 2016).

52. Michelle A. Amazeen et al., "Correcting Political and Consumer Misperceptions: The Effectiveness and Effects of Rating Scale Versus Contextual Correction Formats," *Journalism & Mass Communications Quarterly* 95, no. 1 (Spring 2018): 28–48, https://doi.org/10.1177/1077699016678186.

53. Cass R. Sunstein et al., "How People Update Beliefs about Climate Change: Good News and Bad News," Social Science Research Network, September 2, 2016, https://ssrn.com/abstract=2821919.

54. Edward L. Glaeser and Cass R. Sunstein, "Why Does Balanced News Produce Unbalanced Views?," NBER Working Paper no. 18975, National Bureau of Economic Research, Cambridge, MA, April 2013, https://doi.org/10.3386/w18975.

55. Munro et al., "Biased Assimilation of Sociopolitical Arguments"; and Kahan, "Ideology, Motivated Reasoning, and Cognitive Reflection."

56. David Tewksbury and Julius Matthew Riles, "Polarization as a Function of Citizen Predispositions and Exposure to News on the Internet," *Journal of Broadcasting & Electronic Media* 59, no. 3 (2015): 381–398, https://doi.org/10.1080/08838151.2015.1054996.

57. Boxell, Gentzkow, and Shapiro, "Is the Internet Causing Political Polarization?"

58. I am indebted to Steven Braun, a specialist in scientific and data visualization at Northeastern University, who helped shape and articulate this concept. I am grateful to him for our many discussions leading toward this model.

59. Tim Wu, "Is the First Amendment Obsolete?," Knight First Amendment Institute, Emerging Threats, September 2017, https://knightcolumbia.org/content/tim-wu-first-amendment-obsolete.

60. Cass R. Sunstein, *Republic.com* (Princeton, NJ: Princeton University Press, 2001).

61. Cass R. Sunstein, *#Republic: Divided Democracy in the Age of Social Media* (Princeton, NJ: Princeton University Press, 2017), 253.

62. Ibid.

63. Kathleen Hall Jamieson, "Messages, Micro-Targeting, and New Media Technologies," *Forum* 11, no. 3 (October 2013): 429–435, https://repository.upenn.edu/cgi/viewcontent.cgi?article=1363&context=asc_papers.

64. Jeff Jarvis, "Our Problem Isn't 'Fake News.' Our Problems Are Trust and Manipulation," Medium, June 12, 2017, https://medium.com/whither-news/our-problem-isnt-fake-news-our-problems-are-trust-and-manipulation-5bfbcd716440.

65. Pew Research Center, "Trust Levels of News Sources by Ideological Group," Pew Research Center, October 20, 2014, http://www.journalism.org/2014/10/21/political-polarization-media-habits/pj_14-10-21_mediapolarization-01/.

66. F. A. Hayek, "The Use of Knowledge in Society," *American Economic Review* 35, no. 4 (September 1945): 519–530, http://home.uchicago.edu/~vlima/courses/econ200/spring01/hayek.pdf.

67. Katherine Mangu-Ward, "Wikipedia and Beyond: Jimmy Wales' Sprawling Vision," *Reason*, June 2007, http://reason.com/archives/2007/05/30/wikipedia-and -beyond.

68. Sloman and Fernbach, *The Knowledge Illusion*, 15.

69. Ibid.

70. J. C. R. Licklider, The Council on Library Resources, and Bolt, Beranek, and Newman, *Libraries of the Future* (Cambridge, MA: MIT Press, 1965), https://books. google.com/books/about/Libraries_of_the_future.html?id=PFNpAAAAIAAJ.

71. Bruce R. Schatz, "Information Retrieval in Digital Libraries: Bringing Search to the Net," *Science* 275, (January 17, 1997): 327–334, http://citeseerx.ist.psu.edu/ viewdoc/download?doi=10.1.1.24.3842&rep=rep1&type=pdf.

72. Licklider, The Council on Library Resources, and Bolt, Beranek, and Newman, *Libraries of the Future*, 192.

73. Portions of this section come from Alison J. Head and John Wihbey, "The Importance of Truth Workers in an Era of Factual Recession," April 8, 2017, https:// medium.com/@ajhead1/the-importance-of-truth-workers-in-an-era-of-factual -recession-7487fda8eb3b.

74. Alison J. Head, Michele Van Hoeck, and Kirsten Hostetler, "Why Blogs Endure: A Study of Recent College Graduates and Motivations for Blog Readership," *First Monday* 22, no. 10 (October 2, 2017), http://firstmonday.org/ojs/index.php/fm/ article/view/8065/6539.

75. Ibid.

76. Bill Cope and Mary Kalantzis, "The Role of the Internet in Changing Knowledge Ecologies," *Arbor* 185, no. 737 (May–June 2009): 521–530, https://doi.org/10.3989/ arbor.2009.i737.309.

77. Yochai Benkler, *The Wealth of Networks: How Social Production Transforms Markets and Freedom* (New Haven, CT: Yale University Press, 2006), 12–13.

78. W. A. Gamson and A. Modigliani, "The Changing Culture of Affirmative Action," in *Research in Political Sociology*, ed. R. G. Braungart and M. M. Braungart (Greenwich, CT: JAI Press, 1987), 137–177; and Leighton Walter Kille, "Research Chat: Social Scientist Dietram A. Scheufele on Framing and Political communication," Journalist's Resource, Shorenstein Center on Media, Politics and Public Policy, updated November 19, 2013, https://journalistsresource.org/tip-sheets/research/ research-chat-dietram-scheufele-framing-political-communication.

79. Matt Ford, "Can Bipartisanship End Mass Incarceration?," *Atlantic*, February 25, 2015, https://www.theatlantic.com/politics/archive/2015/02/can-bipartisanship-end -mass-incarceration/386012/.

80. Patterson, *Informing the News*, 22–23.

81. Patrick Sharkey, Gerard Torrats-Espinosa, and Delaram Takyar, "Community and the Crime Decline: The Causal Effect of Local Nonprofits on Violent Crime," *American Sociological Review* 82, no. 6 (October 25, 2017): 1214–1240, http://journals .sagepub.com/doi/abs/10.1177/0003122417736289.

82. Robert D. Crutchfield and Gregory A. Weeks, "The Effects of Mass Incarceration on Communities of Color," *Issues in Science and Technology* 32, no. 1 (Fall 2015), http://issues.org/32-1/the-effects-of-mass-incarceration-on-communities-of-color.

83. Walter C. Dean and Atiba Pertilla, "I-Teams and 'Eye Candy': The Reality of Local TV News," in *We Interrupt This Newscast: How to Improve Local News and Win Ratings, Too*, ed. Tom Rosenstiel et al. (New York: Cambridge University Press, 2007), 30–50, https://doi.org/10.1080/10584600801985755.

84. Storybench, "Reinventing Local TV News Project," Northeastern University School of Journalism, accessed June 19, 2018, http://www.storybench.org/category/tvnews/.

85. Travis L. Dixon, "Good Guys Are Still Always in White? Positive Change and Continued Misrepresentation of Race and Crime on Local Television News," *Communication Research* 44, no. 6 (April 5, 2015): 775–792, http://journals.sagepub.com/doi/abs/10.1177/0093650215579223.

86. Naomi Oreskes and Erik M. Conway, *Merchants of Doubt: How a Handful of Scientists Obscured the Truth on Issues from Tobacco Smoke to Global Warming* (New York: Bloomsbury Press, 2011).

87. Jerome Bruner, "The Narrative Construction of Reality," *Critical Inquiry* 18, no. 1 (Autumn 1991): 1–21, https://doi.org/10.1086/448619.

4   Understanding Media through Network Science

1. Stephen P. Borgatti et al., "Network Analysis in the Social Sciences," *Science* 323, no. 892 (February 13, 2009): 892–895, https://doi.org/10.1126/science.1165821.

2. Albert-László Barabási, "Scale-Free Networks: A Decade and Beyond," *Science* 325, no. 5939 (July 2009): 412–413, https://doi.org/10.1126/science.1173299.

3. Robert K. Merton, "The Matthew Effect in Science, II: Cumulative Advantage and the Symbolism of Intellectual Property," *Isis* 79, no. 4 (December 1988): 606–623, https://doi.org/10.1086/354848.

4. Marc A. Smith et al., "Mapping Twitter Topic Networks: From Polarized Crowds to Community Clusters," Pew Research Center, February 20, 2014, http://www.pewinternet.org/2014/02/20/mapping-twitter-topic-networks-from-polarized-crowds-to-community-clusters/; Itai Himelboim et al., "Classifying Twitter Topic-Networks Using Social Network Analysis," *Social Media and Society* 3, no. 1 (February 1, 2017), https://doi.org/10.1177/2056305117691545.

5. Felippe Rodrigues, "NodeXL's Marc Smith on How Mapping Virtual Crowd Networks Is Relevant to Journalism," Storybench, November 27, 2017, http://www.storybench.org/marc-smith-mapping-social-networks/.

6. Linton Freeman, "A Set of Measures of Centrality Based on Betweenness," *Sociometry* 40 (1977): 35–41.

7. "The Times Issues Social Media Guidelines for the Newsroom," *New York Times*, October 13, 2017, https://www.nytimes.com/2017/10/13/reader-center/social-media-guidelines.html.

8. John Wihbey et al., "Exploring the Ideological Nature of Journalists' Social Networks on Twitter and Associations with News Story Content," paper presented at Data Science + Journalism Workshop, ACM SIGKDD, Halifax, Canada, August 2017, http://lazerlab.net/publication/exploring-ideological-nature-journalists-social-networks-twitter-and-associations-news.

9. See Arthur D. Santana and Toby Hopp, "Tapping into a New Stream of (Personal) Data: Assessing Journalists' Different Use of Social Media," *Journalism & Mass Communication Quarterly* 93, no. 2 (2016): 383–408, https://doi.org/10.1177/1077699016637105; and David H. Weaver and Lars Willnat, "Changes in US Journalism: How Do Journalists Think about Social Media?," *Journalism Practice* 10, no. 7 (2016), 844–855, https://doi.org/10.1080/17512786.2016.1171162.

10. Angela M. Lee, Seth C. Lewis, and Matthew Powers, "Audience Clicks and News Placement: A Study of Time-Lagged Influence in Online Journalism," *Communication Research* 41, no. 4 (2014), 505–530, https://doi.org/10.1177/0093650212467031; and Edson C. Tandoc Jr. and Tim P. Vos, "The Journalist Is Marketing the News: Social Media in the Gatekeeping Process," *Journalism Practice* 10, no. 8 (2016), 950–966, https://doi.org/10.1080/17512786.2015.1087811.

11. Literature in this area dates back very far, to canonical works such as Paul Felix Lazarsfeld, Bernard Berelson, and Hazel Gaudet, *The People's Choice: How the Voter Makes up His Mind in a Presidential Campaign* (New York: Columbia University Press, 1968). Newer studies also bear out the strong nature of social influence, both offline and online. See James Fowler and Nicholas Christakis, "Dynamic Spread of Happiness in a Large Social Network: Longitudinal Analysis of the Framingham Heart Study Social Network," *BMJ (Clinical Research Ed.)* 338, no. 7685 (January 3, 2009): 23–27, https://doi.org/10.1136/bmj.a2338; Adam D. I. Kramer, Jamie E. Guillory, and Jeffrey T. Hancock, "Experimental Evidence of Massive-scale Emotional Contagion through Social Networks," in *Proceedings of the National Academy of Sciences* 111, no. 24 (June 2014): 8788–8790, https://doi.org/10.1073/pnas.1320040111; and Robert M. Bond et al., "A 61-Million-Person Experiment in Social Influence and Political Mobilization," *Nature* 489, no. 7415 (September 2012): 295–298, https://doi.org/10.1038/nature11421.

12. Expressed in more technical terms, an increase in two standard deviations in the level of right-leaning content in a journalist's Twitter network is associated with an increase of 0.35 in production of right-leaning content. In other words, a journalist whose Twitter feed is mostly conservative is likely to produce content that is approximately one-quarter of one standard deviation more conservative than a journalist whose feed is mostly liberal.

13. National Research Council, *Network Science* (Washington, DC: National Academic Press, 2005), https://www.nap.edu/catalog/11516/network-science.

14. Albert-László Barabási, *Linked: The New Science of Networks* (Cambridge, MA: Perseus Books Group, 2002).

15. Barabási, *Network Science*, 24–25.

16. Author interview, Albert-László Barabási, July 18, 2017.

17. P. Erdős and A. Rényi, "On the Evolution of Random Graphs," *Mathematical Institute of the Hungarian Academy of Sciences*, no. 5 (1960): 17–61, https://pdfs .semanticscholar.org/4201/73e087bca0bdb31985e28ff69c60a129c8ef.pdf.

18. Christina Prell, *Social Network Analysis: History, Theory and Methodology* (Thousand Oaks, CA: SAGE Publications Limited, 2012), 39.

19. The two-step flow hypothesis was first proposed in Lazarsfeld, Berelson, and Gaudet, *The People's Choice*. It was confirmed in, among other studies, Paul Felix Lazarsfeld and Elihu Katz, *Personal Influence: The Part Played by People in the Flow of Mass Communications* (New York: Free Press, 1955).

20. Author interview with Duncan Watts, January 8, 2016.

21. Ibid.

22. Duncan J. Watts and Steven H. Strogatz, "Collective Dynamics of 'Small-World' Networks," *Nature* 393 (June 4, 1998): 440–442, https://doi.org/10.1038/30918.

23. Author interview with Duncan J. Watts, January 8, 2016.

24. Author interview with Jonah Peretti, August 28, 2017.

25. See Travis Martin et al., "Exploring Limits to Prediction in Complex Social Systems," in *Proceedings of the 25th International Conference on World Wide Web* (Geneva: International World Wide Web Conferences Steering Committee, 2016), 683–694, https://doi.org/10.1145/2872427.2883001.

26. Author interview with Duncan J. Watts, August 3, 2017.

27. Clive Thompson, "Is the Tipping Point Toast?," Fast Company, February 1, 2008, https://www.fastcompany.com/641124/tipping-point-toast.

28. Tim Wu, *The Attention Merchants: The Epic Scramble to Get Inside Our Heads* (New York: Borzoi Books, 2016), 320.

29. Christopher M. Schroeder, "BuzzFeed Wins the Internet Daily. Here's What Its Boss Thinks Is Next," Recode, December 19, 2016, https://www.recode.net/2016/ 12/19/14010044/buzzfeed-wins-internet-future-of-media-online-social.

30. Ian Burrell, "BuzzFeed's Jonah Peretti: News Publishers Only Have Themselves to Blame for Losing Out to Google and Facebook," *Drum*, June 29, 2017, http://www. thedrum.com/opinion/2017/06/29/buzzfeeds-jonah-peretti-news-publishers-only -have-themselves-blame-losing-out.

31. Janell Sims, "BuzzFeed: The New Newsroom … Is It the Future?," Shorenstein Center on Media, Politics and Public Policy, February 25, 2014, https://shorenstein center.org/ben-smith/.

32. Edson C. Tandoc Jr. and Joy Jenkins, "The Buzzfeedication of Journalism? How Traditional News Organizations Are Talking about a New Entrant to the Journalistic Field Will Surprise You!," *Journalism* 18, no. 4 (December 24, 2015), http://journals .sagepub.com/doi/abs/10.1177/1464884915620269.

33. R. A. Hill and R. I. M. Dunbar, "Social Network Size in Humans," *Human Nature* 14, no. 1 (March 2003): 53–72, https://doi.org/10.1007/s12110-003-1016-y.

34. R. I. M. Dunbar et al., "The Structure of Online Social Networks Mirrors Those in the Offline World," *Social Networks*, no. 43 (October 2015): 39–47, https://doi.org/ 10.1016/j.socnet.2015.04.005.

35. Ibid.

36. General Social Survey, accessed September 1, 2017, http://gss.norc.org/; and Mario Luis Small, "Weak Ties and the Core Discussion Network: Why People Regularly Discuss Important Matters with Unimportant Alters," *Social Networks* 35 (2013): 470–483, https://doi.org/10.1016/j.socnet.2013.05.004.

37. Mark S. Granovetter, "The Strength of Weak Ties," *American Journal of Sociology* 78, no. 6 (May 1973): 1360–1380, https://doi.org/10.1086/225469.

38. Eytan Bakshy et al., "The Role of Social Networks in Information Diffusion."

39. Ibid.

40. Eytan Bakshy et al., "Social Influence in Social Advertising: Evidence from Field Experiments," in *Proceedings of the 13th ACM Conference on Electronic Commerce* (New York: ACM, 2012), 146–161, https://doi.org/10.1145/2229012.2229027.

41. Ethan Zuckerman, *Digital Cosmopolitans in the Age of Connection* (New York: W. W. Norton & Co., 2013), 177–180.

42. Pew Research Internet Project, "Why Most Facebook Users Get More Than They Give," Pew Research Center, February 3, 2012, http://www.pewinternet.org/2012/02/03/why-most-facebook-users-get-more-than-they-give/.

43. Author interview with Mark Granovetter, May 21, 2014; and July 23, 2015.

44. See, for example, L. A. Adamic, and E. Adar, "Friends and Neighbors on the Web," *Social Networks* 25, no. 3 (2003): 211–230, https://doi.org/10.1016/S0378-8733(03)00009-1; and D. Liben-Nowell and J. Kleinberg, "The Link-Prediction Problem for Social Networks," *Journal of the American Society for Information Science and Technology* 58, no. 7 (May 2007): 1019–1031, https://doi.org/10.1002/asi.20591.

45. J. P. Onnela et al., "Structure and Tie Strengths in Mobile Communication Networks," *Proceedings of the National Academy of Sciences of the United States of America* 104, no. 18 (May 1, 2017): 7332–7336, http://www.pnas.org/content/pnas/104/18/7332.full.pdf.

46. John Wihbey, "Does the Secret to Social Networking Lie in the Remote Jungle?," *Boston Globe*, October 4, 2015, https://www.bostonglobe.com/ideas/2015/10/03/does-secret-social-networking-lie-remote-jungle/RnSfIEqW67ZiNHhOmwXIvJ/story.html.

47. Nicholas A. Christakis and James H. Fowler, "The Spread of Obesity in a Large Social Network over 32 Years," *New England Journal of Medicine* 357 (July 26, 2007): 370–379, https://doi.org/10.1056/NEJMsa066082.

48. Kramer, Guillory, and Hancock, "Experimental Evidence of Massive-scale Emotional Contagion."

49. W. Lance Bennett and Alexandra Segerberg, "The Logic of Connective Action," *Information, Communication and Society* 15, no. 5 (April 10, 2012): 739–768, https://doi.org/10.1080/1369118X.2012.670661. These ideas are detailed further in their related book: W. Lance Bennett and Alexandra Segerberg, *The Logic of Connective Action: Digital Media and the Personalization of Contentious Politics* (Cambridge: Cambridge University Press, 2013).

50. Zeynep Tufekçi, *Twitter and Tear Gas: The Power and Fragility of Networked Protest* (New Haven, CT: Yale University Press, 2017).

51. Barry Wellman and Lee Rainie, *Networked: The New Social Operating System* (Cambridge, MA: MIT Press, 2012).

52. Wellman and Rainie, *Networked*, 8.

53. Bond et al., "A 61-Million-Person Experiment."

54. Author interviews with James H. Fowler, March 7, 2014; and September 17, 2015.

55. Facebook Data Science, "Can Cascades Be Predicted?," Facebook, February 28, 2014, https://www.facebook.com/notes/facebook-data-science/can-cascades-be-pre dicted/10152056491448859/.

56. Karine Nahon and Jeff Hemsley, *Going Viral* (Cambridge: Polity Press, 2013), 16.

57. Justin Cheng, "Do Cascades Recur?," paper presented at the International World Wide Web Conference, Montreal, Canada, April 14, 2016, https://arxiv.org/pdf/1602.01107.pdf.

58. Sharad Goel et al., "The Structural Virality of Online Diffusion," *Management Science* 62, no. 1 (July 22, 2015): 180–196, https://doi.org/10.1287/mnsc.2015.2158.

59. Author interview with Sharad Goel, March 25, 2014.

60. Mark Granovetter, "Threshold Models of Collective Behavior," *American Journal of Sociology* 83 (1978): 1420–1443, https://doi.org/10.1086/226707.

61. Lilian Weng, Filippo Menczer, and Yong-Yeol Ahn, "Virality Prediction and Community Structure in Social Networks," *Scientific Reports* 3 (2013): 2522, https://doi.org/10.1038/srep02522.

62. See Barabási, *Network Science*, 405–407; and Duncan J. Watts, *Six Degrees: The Science of a Connected Age* (New York: W. W. Norton & Co., 2004), 239–244.

5   Bias in Network Architectures and Platforms

1. Mike Ananny, *Networked Press Freedom: Creating Infrastructures for a Public Right to Hear* (Cambridge, MA: MIT Press, 2018), 165.

2. Barry Wellman et al., "The Social Affordances of the Internet for Networked Individualism," *Journal of Computer-Mediated Communication* 8, no. 3 (April 2003), https://doi.org/10.1111/j.1083-6101.2003.tb00216.x.

3. Harold Innis, *The Bias of Communication*, 2nd ed. (Toronto: University of Toronto Press, 1999); Larry Lessig, *Code: And Other Laws of Cyberspace Version 2.0*, 2nd ed. (New York: Basic Books, 2006); and Benkler, *The Wealth of Networks*, 16.

4. Kiran Garimella, Ingmar Weber, and Munmun De Choudhury, "Quote RTs on Twitter: Usage of the New Feature for Political Discourse," in *Proceedings of the 8th ACM Conference on Web Science*, March 25, 2016, https://arxiv.org/pdf/1603.07933.pdf.

5. Sarah Frier, "Facebook Reaches 2 Billion Users, Now Wants to Build Friendships," *Bloomberg News*, June 27, 2017, https://www.bloomberg.com/news/articles/2017 -06-27/facebook-reaches-2-billion-users-now-wants-to-build-friendships; and Danny Sullivan, "Google Now Handles at Least 2 Trillion Searches per Year," Search Engine

Land, May 24, 2016, https://searchengineland.com/google-now-handles-2-999 -trillion-searches-per-year-250247.

6. Journalistic monitoring and intervention in comment threads can help. See *The Engaging News Project: Journalist Involvement in Comment Sections*, University of Texas at Austin Center for Media Engagement, September 10, 2014, https://www.demo cracyfund.org/media/uploaded/ENP_Comments_Report.pdf.

7. See Ashley A. Anderson et al., "The 'Nasty Effect:' Online Incivility and Risk Perceptions of Emerging Technologies," *Journal of Computer-Mediated Communication* 19, no. 3 (February 19, 2013): 373–387, https://onlinelibrary.wiley.com/doi/10.1111/ jcc4.12009/abstract; and Eun-Ju Lee, "That's Not the Way It Is: How User-Generated Comments on the News Affect Percieved Media Bias," *Journal of Computer-Mediated Communication* 18, no. 1 (October 10, 2012): 32–45, https://doi.org/10.1111/j.1083 -6101.2012.01597.x.

8. Justin Cheng, Cristian Danescu-Niculescu-Mizil, and Jure Leskovec, "How Community Feedback Shapes User Behavior," in *Proceedings of the Eighth International AAAI Conference on Weblogs and Social Media* (Palo Alto, CA: AAAI Press, 2014), 41–50, https://www.aaai.org/ocs/index.php/ICWSM/ICWSM14/paper/viewFile/8066/ 8104.

9. James Gamble, "Wiring a Continent: The Making of the U.S. Transcontinental Telegraph," published in *The Californian*, 1881, accessed January 20, 2018, http:// www.telegraph-history.org/transcontinental-telegraph/.

10. David Hochfelder, *The Telegraph in America, 1832–1920* (Baltimore: Johns Hopkins University Press, 2012).

11. Ibid., 3.

12. *Oxford Dictionary Online*, s.v. "net-work," accessed August 1, 2017, http://www .oed.com/view/Entry/126342?rskey=lQKY6x&result=1#eid.

13. James Gleick, *The Information* (New York: Pantheon Books, 2011), 151.

14. *Harper's New Monthly Magazine*, "The Telegraph," *Harper's New Monthly Magazine* 47, no. 279 (August 1873): 332–360.

15. James W. Carey, "Technology and Ideology: The Case of the Telegraph," in *Communication as Culture: Essays on Media and Society*, rev. ed. (New York: Routledge, 2009), 155–177.

16. Hochfelder, *The Telegraph in America*.

17. Christopher B. Daly, *Covering America: A Narrative History of a Nation's Journalism* (Amherst: University of Massachusetts Press, 2012).

18. David Kirkpatrick, *The Facebook Effect: The Inside Story of the Company that Is Connecting the World* (New York: Simon & Schuster, 2011).

19. Mark Zuckerberg, lecture, CS50, December 7, 2005, YouTube video, https:// www.youtube.com/watch?v=xFFs9UgOAlE&t=358s.

20. Ibid.

21. Bob Metcalfe, "Metcalfe's Law after 40 Years of Ethernet," in *Computer* 46, no. 12 (2013): 26–31, http://doi.org/10.1109/MC.2013.374.

22. Harry Lewis, "My REAL Contribution to the Birth of Facebook (II)," *Bits and Pieces* (blog), May 19, 2012, http://harry-lewis.blogspot.com/2012/05/my-real-contribution-to-birth-of.html.

23. Peter Dodds, Roby Muhamad, and Duncan Watts, "An Experimental Study of Search in Global Social Networks," *Science* 301, no. 5634 (2003): 827–829, https://doi.org/10.1126/science.1081058.

24. Lars Backstrom et al., "Four Degrees of Separation," paper presented at the Third Annual ACM Web Science Conference, Evanston, IL, June 22–24, 2012, https://doi.org/10.1145/2380718.2380723.

25. Smriti Bhagat et al., "Three and a Half Degrees of Separation," Facebook Research (blog), February 4, 2016, https://research.fb.com/three-and-a-half-degrees-of-separation/.

26. Emily Bell, "Who Owns the News Consumer: Social Media Platforms or Publishers?," *Columbia Journalism Review*, June 21, 2016, https://www.cjr.org/tow_center/platforms_and_publishers_new_research_from_the_tow_center.php.

27. See, for example, Rick Edmonds, "News Media Alliance Seeks Antitrust Exemption to Negotiate a Better Deal with Facebook and Google," Poynter, July 10, 2017, https://www.poynter.org/2017/news-media-alliance-seeks-antitrust-exemption-to-negotiate-a-better-deal-with-facebook-and-google/466126/; and Fidji Simo, "Introducing: The Facebook Journalism Project," *Facebook Journalism Project* (blog), January 11, 2017, https://media.fb.com/2017/01/11/facebook-journalism-project/.

28. Sahil Patel, "Facebook Recruits Its Top Publishers for Exclusive Shows," Digiday, May 25, 2017, https://digiday.com/media/facebook-recruits-top-video-publishers-shows/.

29. Ana Lucía Schmidt et al., "Anatomy of News Consumption on Facebook," *Proceedings of the National Academy of Sciences* 114, no. 12 (March 1, 2017): 3035–3039, https://doi.org/10.1073/pnas.1617052114.

30. Eytan Bakshy, Solomon Messing, and Lada Adamic, "Exposure to Ideologically Diverse News and Opinion on Facebook," in *Science* 348, no. 6239 (June 2015): 1130–1132, https://doi.org/10.1126/science.aaa1160.

31. See Christian Sandvig, "The Facebook 'It's Not Our Fault' Study," *Social Media Collective Research Blog*, May 7, 2015, https://socialmediacollective.org/2015/05/07/the-facebook-its-not-our-fault-study/; and Zeynep Tufekci, "How Facebook's Algorithm Suppresses Content Diversity (Modestly) and How the Newsfeed Rules Your Clicks," *The Message* (blog), May 7, 2015, https://medium.com/message/how-facebook-s-algorithm-suppresses-content-diversity-modestly-how-the-newsfeed-rules-the-clicks-b5f8a4bb7bab.

32. David Lazer, "The Rise of the Social Algorithm," *Science* 1 (May 7, 2015): 1090–1091, https://doi.org/10.1126/science.aab1422.

33. Kurt Wagner, "Facebook Found a New Way to Identify Spam and False News Articles in Your News Feed," Recode, June 30, 2017, https://www.recode.net/2017/6/30/15896544/facebook-fake-news-feed-algorithm-update-spam.

34. For example, see Bond et al., "A 61-Million-Person Experiment in Social Influence and Political Mobilization"; Kramer, Guillory, and Hancock, "Experimental

Evidence of Massive-Scale Emotional Contagion through Social Networks"; Moira Burke and Robert Kraut, "Growing Closer on Facebook: Changes in Tie Strength Through Social Network Site Use," in *Proceedings of the SIGCHI Conference on Human Factors in Computing Systems* (New York: ACM, 2014), 4187–4196; and Eytan Bakshy et al., "The Role of Social Networks in Information Diffusion," in *Proceedings of the 21st International Conference on World Wide Web* (New York: ACM, 2012), 519–528, https://doi.org/10.1145/2187836.2187907.

35. Google, "About," accessed August 1, 2017,https://www.google.com/intl/en/about/.

36. Steven Levy, *In the Plex: How Google Thinks, Works, and Shapes Our Lives* (New York: Simon & Schuster, 2011), 24.

37. John Battelle, *The Search: How Google and Its Rivals Rewrote the Rules of Business and Transformed Our Culture* (New York: Penguin, 2005), 73.

38. Stanford InfoLab, "The PageRank Citation Ranking: Bringing Order to the Web," technical report, Stanford InfoLab, 1998, http://ilpubs.stanford.edu:8090/422/1/1999-66.pdf.

39. Julien Chokkattu, "New Google Algorithm Lowers Search Rankings for Holocaust Denial Sites," *Digital Trends*, December 25, 2016, https://www.digitaltrends.com/web/google-search-holocaust/.

40. Eli Pariser, *The Filter Bubble: How The New Personalized Web Is Changing What We Read and How We Think* (New York: Penguin Books Limited, 2011); and Siva Vaidhyanathan, *The Googlization of Everything (And Why We Should Worry)* (Berkeley: University of California Press, 2011).

41. Aniko Hannak et al., "Measuring Personalization of Web Search," in *Proceedings of the 22nd International Conference on World Wide Web* (New York: ACM, 2013), 527–538, https://dl.acm.org/citation.cfm?id=2488435.

42. Mimi Underwood, "Updating Our Search Quality Rating Guidelines," *Google Webmaster Central Blog*, November 19, 2015, https://webmasters.googleblog.com/2015/11/updating-our-search-quality-rating.html.

43. John Huey, Martin Nisenholtz, and Paul Sagan, "Riptide: What Really Happened to the News Business," Shorenstein Center on Media, Politics and Public Policy, September 8, 2013, https://shorensteincenter.org/2013/09/d81-riptide/.

44. Nicholas Carr, "Is Google Making Us Stupid? What the Internet Is Doing to Our Brains," *Atlantic*, July/August 2008, https://www.theatlantic.com/magazine/archive/2008/07/is-google-making-us-stupid/306868/.

45. Lucia Moses, "'Never More Accurate': Jack Shafer on Journalism in 2014," Digiday, December 18, 2014, https://digiday.com/media/jack-shafer-journalism-2014-never-accurate/.

6   Data, Artificial Intelligence, and the News Future

1. See "Live Presidential Forecast," *New York Times*, November 9, 2016, https://www.nytimes.com/elections/forecast/president.

2. David T. Z. Mindich, *Just the Facts: How "Objectivity" Came to Define American Journalism* (New York: New York University Press, 1998), 65. Other scholars put the date of emergence for the inverted period slightly later. See Marcus Errico, "The Evolution of the Summary News Lead," *Media History Monographs* 1, no. 1 (1997), https://www.scripps.ohiou.edu/mediahistory/mhmjour1-1.htm.

3. Nielsen, "Digital News as Forms of Knowledge," 108.

4. See, for example, Carl Benedikt Frey and Michael A. Osborne, "The Future of Employment: How Susceptible Are Jobs to Computerisation?," Oxford Martin Programme on Technology and Employment, September 17, 2013, https://www.oxfordmartin.ox.ac.uk/downloads/academic/The_Future_of_Employment.pdf.

5. Tim Adams, "And the Pulitzer Goes to … a Computer," *Guardian*, June 28, 2015, https://www.theguardian.com/technology/2015/jun/28/computer-writing-journalism-artificial-intelligence.

6. Pedro Domingos, *The Master Algorithm: How the Quest for the Ultimate Learning Machine Will Remake the World* (New York: Basic Books, 2015), 8.

7. "Artificial Intelligence: Practice and Implications for Journalism," conference event streamed by Columbia Journalism School, June 13, 2017, YouTube video, https://www.youtube.com/watch?v=vwaJKJ-zya0&t=1339s.

8. Amy Webb, "2018 Tech Trends for Journalism and Media," Future Today Institute, October 2017, https://futuretodayinstitute.com/2018-tech-trends-for-journalism-and-media/.

9. Ricardo Bilton, "Reuters Built Its Own Algorithmic Prediction Tool to Help It Spot (and Verify) Breaking News on Twitter," Nieman Lab, November 30, 2016, http://www.niemanlab.org/2016/11/reuters-built-its-own-algorithmic-prediction-tool-to-help-it-spot-and-verify-breaking-news-on-twitter/.

10. Francesco Marconi, *How Artificial Intelligence Will Impact Journalism*. Associated Press, April 5, 2017, https://insights.ap.org/industry-trends/report-how-artificial-intelligence-will-impact-journalism.

11. Jonathan Stray, "The Age of the Cyborg," *Columbia Journalism Review*, Fall/Winter 2016, https://www.cjr.org/analysis/cyborg_virtual_reality_reuters_tracer.php.

12. See Armineh Nourbakhsh et al., "'Breaking' Disasters: Predicting and Characterizing the Global News Value of Natural and Man-Made Disasters," paper presented at SIGKDD, Data Science + Journalism Workshop, Halifax, Canada, August 12–17, 2017, https://www.researchgate.net/publication/318468631_Breaking_Disasters_Predicting_and_Characterizing_the_Global_News_Value_of_Natural_and_Man-made_Disasters; Quanzhi Li et al., "RealTime Novel Event Detection from Social Media," in *2017 IEEE 33rd International Conference on Data Engineering (ICDE)* (Piscataway, NJ: IEEE, 2017), 1129–1139, https://doi.org/10.1109/ICDE.2017.157; Xiaomo Liu et al. "Reuters Tracer: A Large Scale System of Detecting & Verifying RealTime News Events from Twitter." *Proceedings of the 25th ACM International on Conference on Information and Knowledge Management ACM* (2017): 207–216, https://doi.org/10.1145/2983323.2983363; and Xiaomo Liu et al., "Real-time Rumor Debunking on

Twitter," in *Proceedings of the 24th ACM International Conference on Information and Knowledge Management* (New York: ACM, 2015), 1867–1870.

13. "The Making of Reuters News Tracer," Thomson Reuters, April 25, 2017, https://blogs.thomsonreuters.com/answerson/making-reuters-news-tracer/.

14. This section on AI and machine learning benefitted from insights synthesized in Mark Hansen et al., "Artificial Intelligence: Practice and Implications for Journalism," Tow Center for Digital Journalism, Columbia University, September 2017, https://academiccommons.columbia.edu/catalog/ac:gf1vhhmgs8.

15. "How the Doctors and Sex Abuse Project Came About," *Atlanta Journal-Constitution*, 2016, http://doctors.ajc.com/about_this_investigation/.

16. Susannah Nesmith, "How a Regional Newspaper Pulled Off a National Investigation into Sexual Abuse by Doctors," *Columbia Journalism Review*, July 11, 2016, https://www.cjr.org/united_states_project/atlanta_journal_constitution_doctor_sex_abuse_investigation.php.

17. Ben Poston and Anthony Pesce, "How We Reported This Story," *Los Angeles Times*, October 15, 2015, http://www.latimes.com/local/cityhall/la-me-crime-stats-side-20151015-story.html.

18. Keith Kirkpatrick, "Putting the Data Science into Journalism," *Communications of the ACM* 58, no. 5 (May 2015): 15–17, https://doi.org/10.1145/274284. See also Meredith Broussard, *Artificial Unintelligence: How Computers Misunderstand the World* (Cambridge, MA: MIT Press, 2018).

19. See Executive Office of the President, "Try This at Home," White House Office of Science and Technology, January 2017, https://www.whitehouse.gov/sites/whitehouse.gov/files/images/Try%20This%20At%20Home-01-2017.pdf; and Leonard Downie Jr., "The Obama Administration and the Press," Committee to Protect Journalists, October 10, 2013, https://cpj.org/reports/2013/10/obama-and-the-press-us-leaks-surveillance-post-911.php.

20. David Cuillier, *Forecasting Freedom of Information: Why it Faces Problems—and How Experts Say They Could Be Solved* (Miami, FL: John S. and James L. Knight Foundation, 2017).

21. Michael Morisy, "Under Trump's First 100 Days, FOIA a Little Slower while Open Data Takes a Hit," MuckRock, April 28, 2017, https://www.muckrock.com/news/archives/2017/apr/28/under-trumps-first-100-days-foia-little-slower-whi/.

22. Daniel Kreiss, "Beyond Administrative Journalism: Civic Skepticism and the Crisis in Journalism," in *The Crisis of Journalism Reconsidered: Democratic Culture, Professional Codes, Digital Future*, ed. Jeffrey C. Alexander, Elizabeth Butler Breese, and Maria Luengo (Cambridge: Cambridge University Press, 2016), 59.

23. Michael Hudson, "Panama Papers Wins Pulitzer Prize," International Consortium of Investigative Journalists, April 10, 2017, https://www.icij.org/blog/2017/04/panama-papers-wins-pulitzer-prize/.

24. Michael Schudson, *The Rise of the Right to Know: Politics and the Culture of Transparency, 1945–1975* (Cambridge, MA: Harvard University Press, 2015).

25. Archon Fung et al., "Transparency Policies: Two Possible Futures," *Taubman Center Policy Briefs* 1 (May 2007): 1–7, http://www.transparencypolicy.net/assets/two %20possible%20futures.pdf.

26. Schudson, *Rise of the Right to Know*.

27. Beth A. Rosenson, "Against Their Apparent Self-Interest: The Authorization of Independent State Legislative Ethics Commissions," *State Politics and Policy Quarterly* 3, no. 1 (Spring 2003): 42–65.

28. This section draws on work previously published in John Wihbey and Michael Beaudet, "State-Level Policies for Personal Financial Disclosure: Exploring the Potential for Public Engagement on Conflict-of-Interest Issues" (paper presented at the Association for Education in Journalism and Mass Communication, Law and Policy Division, Chicago, Illinois, August 2017); John Wihbey and Michael Beaudet, "Why It's So Hard to See Politicians' Financial Data," *New York Times*, October 4, 2016, https://www.nytimes.com/2016/10/04/opinion/why-its-so-hard-to-see-politicians -financial-data.html; and John Wihbey, Michael Beaudet, and Pedro Miguel Cruz, "There Are Huge Holes in How the U.S. States Investigate Politicians' Conflicts of Interest," *Washington Post*, January 12, 2017, https://www.washingtonpost.com/ news/monkey-cage/wp/2017/01/12/how-do-states-investigate-officials-potential -conflicts-of-interest-we-checked/.

29. Richard J. Tofel, "Why Corruption Grows in Our States: Fewer Reporters and Remote State Capitals," *Daily Beast*, May 24, 2010, https://www.thedailybeast.com/ why-corruption-grows-in-our-states-fewer-reporters-and-remote-state-capitals.

30. Louis D. Brandeis, *Other People's Money, and How the Bankers Use It* (New York: Frederick A. Stokes, 1914).

31. See Mike Beaudet et al., "The State Financial Disclosure Project," Northeastern University College of Arts, Media and Design, 2016, https://web.northeastern.edu/ disclosure-project/.

7   Journalism's New Approach to Knowledge

1. Lippmann, *Liberty and the News,* 8.

2. Ibid., 11.

3. Ibid., 5.

4. Michael Schudson, "The 'Lippmann-Dewey Debate' and the Invention of Walter Lippmann as an Anti-Democrat 1986–1996," *International Journal of Communication* 2 (September 2008): 1032–1042.

5. Ibid.

6. For example, see Ethan Zuckerman, "New Media, New Civics?," *Policy and Internet* 6, no. 2 (June 2014): 151–168, https://doi.org/10.1002/1944-2866.POI360.

7. John Dewey, "Public Opinion," *New Republic*, May 3, 1922.

8. Ibid.

9. Amy Gutmann and Dennis Thompson, *Why Deliberative Democracy?* (Princeton, NJ: Princeton University Press, 2004).

10. Ettema, "Journalism as Reason-Giving."

11. See Neena Satija et al., "Hell and High Water," ProPublica, March 3, 2016, https://projects.propublica.org/houston/; and Al Shaw and Jeff Larson, "How We Made Hell and High Water," ProPublica, March 3, 2016, https://www.propublica.org/nerds/how-we-made-hell-and-high-water.

12. "Fatal Force," database, *Washington Post*, accessed June 19, 2018, https://www.washingtonpost.com/policeshootings/; and "The Counted," *Guardian*, accessed March 15, 2018, https://www.theguardian.com/us-news/ng-interactive/2015/jun/01/the-counted-police-killings-us-database.

13. Aaron C. Davis and Wesley Lowery, "FBI Director Calls Lack of Data on Police Shootings 'Ridiculous,' 'Embarrassing,'" *Washington Post*, October 7, 2015, https://www.washingtonpost.com/national/fbi-director-calls-lack-of-data-on-police-shootings-ridiculous-embarrassing/2015/10/07/c0ebaf7a-6d16-11e5-b31c-d80d62b53e28_story.html.

14. Gabriel Dance and Tom Meagher, "Crime in Context," The Marshall Project, August 18, 2016, https://www.themarshallproject.org/2016/08/18/crime-in-context.

15. Julia Angwin et al., "Machine Bias," ProPublica, May 23, 2016, https://www.propublica.org/article/machine-bias-risk-assessments-in-criminal-sentencing.

16. Nicholas Diakopoulos, "Algorithmic Accountability: Journalistic Investigation of Computational Power Structures," *Digital Journalism*, no. 3 (November 7, 2014): 398–415, https://doi.org/10.1080/21670811.2014.976411.

17. Jeff Larson et al., "How We Analyzed the COMPAS Recidivism Algorithm," ProPublica, May 23, 2016, http://www.propublica.org/article/how-we-analyzed-the-compas-recidivism-algorithm/. For a look at the parallel research literature, which the journalists explicitly cite and draw on, see Jennifer L. Skeem and Christopher T. Lowenkamp, "Risk, Race, & Recidivism: Predictive Bias and Disparate Impact," Social Science Research Network, June 15, 2016, https://doi.org/10.2139/ssrn.2687339.

18. Sam Corbett-Davies et al., "A Computer Program Used for Bail and Sentencing Decisions Was Labeled Biased against Blacks. It's Actually Not that Clear," *Washington Post*, October 17, 2016, https://www.washingtonpost.com/news/monkey-cage/wp/2016/10/17/can-an-algorithm-be-racist-our-analysis-is-more-cautious-than-propublicas/?utm_term=.89099c54847f.

19. Lincoln Steffens, *The Autobiography of Lincoln Steffens* (New York: Harcourt, Brace, and World, 1936). A useful account of this can also be found in Michael Schudson, *The Sociology of News* (New York: W. W. Norton & Company, 2003), 1–4.

20. Walter Lippmann, *Public Opinion* (New York: Free Press Paperbacks, 1997), 3–9.

21. Patterson, *Informing the News*, 65–79.

22. Wolfgang Donsbach, "Journalism as the New Knowledge Profession and Consequences for Journalism Education," *Journalism* 15, no. 6 (July 2, 2013): 661–677, https://doi.org/:10.1177/1464884913491347.

23. Matthew Nisbet and Declan Fahy, "The Need for Knowledge-Based Journalism in Politicized Science Debates," *Annals of the American Academy of Political and Social Science* 658, no. 1 (February 8, 2015): 223–234, https://doi.org/10.1177/0002716214559887.

24. Mitchell Stephens, *Beyond News: The Future of Journalism* (New York: Columbia University Press, 2014), 177–178.

25. Kevin G. Barnhurst, "The Problem of Realist Events in American Journalism," *Media and Communication* 2, no. 2, (October 2014): 84–95, https://doi.org/10.17645/mac.v2i2.159.

26. Kevin G. Barnhurst and Diana Mutz, "American Journalism and the Decline in Event-Centered Reporting," *Journal of Communication* 47, no. 4 (1997): 27–53.

27. Katherine Fink and Michael Schudson, "The Rise of Contextual Journalism, 1950s–2000s," *Journalism* 15, no. 1 (February 17, 2013): 3–20, https://doi.org/10.1177/1464884913479015.

28. Susan E. McGregor, "A Brief History of Computer Assisted Reporting," *Tow Center for Digital Journalism* (blog), March 18, 2013, https://towcenter.org/a-brief-history-of-computer-assisted-reporting/.

29. Stephen K. Doig, "Reporting with the Tools of Social Science," *Nieman Reports*, March 15, 2008, https://niemanreports.org/articles/reporting-with-the-tools-of-social-science/.

30. Lars Willnat and David H. Weaver, "The American Journalist in the Digital Age: Key Findings," School of Journalism, Indiana University, 2014, http://archive.news.indiana.edu/releases/iu/2014/05/2013-american-journalist-key-findings.pdf.

31. Dewey, *Public and Its Problems*, 180–181.

32. Philip Meyer, *Precision Journalism: A Reporter's Guide to Social Science Methods*, 4th ed. (Lanham, MD: Rowan & Littlefield, 2002).

33. Philip Meyer, "The Next Journalism's Objective Reporting," *Nieman Reports*, December 15, 2004, https://niemanreports.org/articles/the-next-journalisms-objective-reporting/.

34. Patterson, *Informing the News*, 65.

35. Donsbach, "Journalism as the New Knowledge Profession."

36. G. Pascal Zachary, "To Prepare 21st-Century Journalists, Help Students Become Experts," *Chronicle of Higher Education*, December 1, 2014, http://www.chronicle.com/article/To-Prepare-21st-Century/150267/.

37. Picard, "Deficient Tutelage."

38. Serena Carpenter, "An Application of the Theory of Expertise: Teaching Broad and Skill Knowledge Areas to Prepare Journalists for Change," *Journalism & Mass Communication Educator* 64 (September 1, 2009): 287–304, https://doi.org/10.1177/107769580906400305.

39. Michael Ryan and Les Switzer, "Balancing Arts and Sciences, Skills, and Conceptual Content," *Journalism & Mass Communication Educator* 56, no. 2 (June 1, 2001): 55–68, https://doi.org/10.1177/107769580105600205.

40. Jean Folkerts, John Maxwell Hamilton, and Nicholas Lemann, "Educating Journalists: A New Plea for the University Tradition," Columbia University Graduate School of Journalism, October 2013, https://www.journalism.columbia.edu/system/documents/785/original/75881_JSchool_Educating_Journalists-PPG_V2-16.pdf.

41. G. Stuart Adam, "The Education of Journalists," *Journalism* 2, no. 3 (December 1, 2001): 315–339, https://doi.org/10.1177/146488490100200309.

42. Alison J. Head and John Wihbey, "At Sea in a Deluge of Data," *Chronicle of Higher Education*, July 7, 2014, http://www.chronicle.com/article/At-Sea-in-a-Deluge-of -Data/147477/.

43. Patterson, *Informing the News*, 103.

44. Nicholas Lemann, "The Journalistic Method: Five Principles for Blending Analysis and Narrative," Journalist's Resource, Shorenstein Center on Media, Politics and Public Policy, updated April 8, 2016, https://journalistsresource.org/tip-sheets/ journalistic-method-tip-sheet-blending-analysis-narrative.

45. This graphic draws in part on ibid.

46. Patterson, *Informing the News*, 66.

47. Jonathan Stray, "Network Analysis in Journalism: Practices and Possibilities" (paper presented at Data Science + Journalism Workshop, ACM SIGKDD in Halifax, Canada, August 2017), https://drive.google.com/file/d/0B8CcT_0LwJ8QMzFjTWxLS FVkVTg/view.

48. Per Jonathan Stray's research, see Tricia Romano, "In Seattle Art World, Women Run the Show," *Seattle Times*, June 18, 2016, https://www.seattletimes.com/ entertainment/visual-arts/in-seattle-art-world-women-run-the-show/; Zachary Sampson and Lisa Gartner, "Hot Wheels," *Tampa Bay Times*, April 26, 2017, http://www .tampabay.com/projects/2017/investigations/florida-pinellas-auto-theft-kids-hot- wheels/car-theft-epidemic/; and Mar Cabra, "How the ICIJ Used Neo4j to Unravel the Panama Papers," *Neo4j Blog*, May 12, 2016, https://neo4j.com/blog/icij-neo4j -unravel-panama-papers/.

49. Stray, "Network Analysis in Journalism."

50. See, for example, Liliana Bounegru et al., "Narrating Networks," *Digital Journalism* 5, no. 6 (2017): 699–730, https://doi.org/10.1080/21670811.2016.1186497.

51. The survey was conducted through the Journalist's Resource project (journalistsresource.org). Begun in 2009, it was originally funded by the Carnegie Corporation of New York and the Knight Foundation.

52. The analysis presented in this section is based on two published papers: John Wihbey, "Journalists' Use of Knowledge in an Online World: Examining Reporting Habits, Sourcing Practices and Institutional Norms," *Journalism Practice* 11, no. 10 (November 3, 2016): 1267–1282, https://doi.org/10.1080/17512786.2016.1249004; and John Wihbey and Mark Coddington, "Knowing the Numbers: Assessing Attitudes among Journalists and Educators about Using and Interpreting Data, Statistics, and Research," *#ISOJ Journal* 7, no. 1 (April 2017), http://isoj.org/research/knowing -the-numbers-assessing-attitudes-among-journalists-and-educators-about-using-and -interpreting-data-statistics-and-research/. It is also based on findings from an unpublished 2016 survey of health science journalists conducted by the Shorenstein Center on Media, Politics and Public Policy. Survey results may be accessed at https://repository.library.northeastern.edu/collections/neu:m0417171j.

53. Patterson, *Informing the News*, 74.

54. Holly Yettick, "One Small Droplet: News Media Coverage of Peer-Reviewed and University-Based Education Research and Academic Expertise," *Education Researcher* 44, no. 3 (2015): 173–184, https://doi.org/10.3102/0013189X15574903.

55. Scott R. Maier, "Accuracy Matters: A Cross-Market Assessment of Newspaper Error and Credibility," *Journalism & Mass Communication Quarterly* 82, no. 3 (2005): 533–551, https://doi.org/10.1177/107769900508200304.

56. See, for example, Robert J. Haiman, "Best Practices for Newspaper Journalists," Free Press/Fair Press Project, Freedom Forum, Arlington, VA, September 3, 2002; and "Media Use and Evaluation," Gallup News, March 14, 2017, http://news.gallup .com/poll/1663/media-use-evaluation.aspx.

57. "Confidence in Institutions: Trends in Americans' Attitudes toward Government, Media, and Business," Associated Press and NORC, March 2015, http://www. apnorc.org/projects/Pages/confidence-in-institutions-trends-in-americans-attitudes- toward-government-media-and-business.aspx.

58. See Angela Phillips, "Transparency and the New Ethics of Journalism," *Journalism Practice* 4, no. 3 (July 8, 2010): 373–382, https://doi.org/10.1080/17512781003 642972; and Tom Rosenstiel, "Why 'Be Transparent' Has Replaced 'Act Independently' as a Guiding Journalism Principle," Poynter, September 16, 2013, https:// www.poynter.org/news/why-be-transparent-has-replaced-act-independently-guiding -journalism-principle.

59. Nate Silver, "The Media Has a Probability Problem," FiveThirtyEight, September 21, 2017, https://fivethirtyeight.com/features/the-media-has-a-probability-problem/.

60. Thomas Jefferson, letter to Edward Carrington, January 16, 1787, http://tjrs. monticello.org/letter/1289.

61. John Norvell, letter to Thomas Jefferson, March 5, 1809, https://founders .archives.gov/?q=Correspondent%3A%22Norvell%2C%20John%22%20 Correspondent%3A%22Jefferson%2C%20Thomas%22&s=1111311111&r=3.

62. Thomas Jefferson, letter to John Norvell, June 11, 1807, https://founders .archives.gov/documents/Jefferson/99-01-02-5737.

63. Lindsey Bever, "Memo to Donald Trump: Thomas Jefferson Invented Hating the Media," *Washington Post*, February 18, 2017, https://www.washingtonpost.com/ news/the-fix/wp/2017/02/17/trumps-war-with-the-media-isnt-new-thomas -jefferson-railed-about-newspaper-lies-too/.

64. Jefferson, letter to Norvell.

65. For example, see Jakob D. Jensen, "Scientific Uncertainty in News Coverage of Cancer Research: Effects of Hedging on Scientists' and Journalists' Credibility," *Human Communication Research* 34, no. 3 (2008): 347–369, https://doi.org/ 10.1111/j.1468-2958.2008.00324.x; and Sharon M. Friedman, Sharon Dunwoody, and Carol L. Rogers, *Communicating Uncertainty: Media Coverage of New and Controversial Science* (New York: Routledge, 1999).

66. Victor Cohn, Lewis Cope, and Deborah Cohn Runkle, *News and Numbers: A Writer's Guide to Statistics*, 3rd ed. (Hoboken, NJ: Wiley-Blackwell, 2011).

67. Andrew Gelman, "Science Journalism and the Art of Expressing Uncertainty," The Symposium, August 4, 2013, http://www.symposium-magazine.com/science -journalism-and-the-art-of-expressing-uncertainty/.

68. Aleszu Bajak, "Should Newspapers Be Adding Confidence Intervals to Their Graphics?," Storybench, May 19, 2017, http://www.storybench.org/newspapers -adding-confidence-intervals-graphics/.

69. Enrico Bertini and Moritz Stefaner, "Uncertainty and Trumpery with Alberto Cairo," *Data Stories*, March 21, 2017, podcast audio, http://datastori.es/94 -uncertainty-and-trumpery-with-alberto-cairo/.

70. Cary Funk and Sara Kehaulani Goo, "A Look at What the Public Knows and Does Not Know about Science," Pew Research Center, September 10, 2015, http://www .pewinternet.org/2015/09/10/what-the-public-knows-and-does-not-know-about -science/.

71. Susanne Tak, Alexander Toet and Jan van Erp, "The Perception of Visual Uncertainty Representation by Non-experts," *IEEE Transactions on Visualization and Computer Graphics* 20, no. 6 (June 2014): 935–943, https://ieeexplore.ieee.org/abstract/ document/6654171/; and David Spiegelhalter, Mike Pearson, and Ian Short, "Visualizing Uncertainty about the Future," *Science* 333, no. 6048 (September 9, 2011): 1393–1400, http://science.sciencemag.org/content/333/6048/1393.full.

72. Charles Weiss, "Expressing Scientific Uncertainty," *Law, Probability and Risk* 2, no. 1 (March 2003), 25–46.

73. See John Wihbey, "Guide to Critical Thinking, Research, Data and Theory: Overview for Journalists," Journalist's Resource, Shorenstein Center on Media, Politics and Public Policy, updated March 6, 2015, https://journalistsresource.org/tip-sheets/ research/guide-academic-methods-critical-thinking-theory-overview-journalists.

74. Lemann, "Journalistic Method."

8 Questions for Engaged Journalism

1. Jake Batsell, *Engaged Journalism: Connecting with Digitally Empowered News Audiences* (New York: Columbia University Press, 2015), 9–12.

2. Dan Kennedy, "Tracing the Links between Civic Engagement and the Revival of Local Journalism," Nieman Lab, June 4, 2013, http://www.niemanlab.org/2013/06/ tracing-the-links-between-civic-engagement-and-the-revival-of-local-journalism/; and Dan Kennedy, *The Wired City: Reimagining Journalism and Civic Life in the Postnewspaper Age* (Amherst: University of Massachusetts Press, 2013).

3. Morris Janowitz, "Professional Models in Journalism: The Gatekeeper and the Advocate," *Journalism & Mass Communication Quarterly* 52, no. 4 (December 1, 1975): 618–626, https://doi.org/10.1177/107769907505200402.

4. See, for example, Robert M. Entman and Andrew Rojecki, *The Black Image in the White Mind: Media and Race in America* (Chicago: University of Chicago Press, 2001).

5. Jay Rosen, "The People Formerly Known as the Audience," Press Think, June 27, 2006, http://archive.pressthink.org/2006/06/27/ppl_frmr.html.

6. Kovach and Rosenstiel, *Elements of Journalism*; and Bill Kovach and Tom Rosenstiel, *Blur: How to Know What's True in the Age of Information Overload* (New York: Bloomberg, 2011).

7. Lucia Moses, "As Audience Development Grows, Publishers Question Who Should Own It," Digiday, June 14, 2017, https://digiday.com/media/audience-development -grows-publishers-question/.

8. Melissa Tully et al., "Case Study Shows Disconnect on Civic Journalism's Role," *Newspaper Research Journal* 38, no. 4 (Fall 2017): 484–496, https://doi.org/ 10.1177/0739532917739881.

9. Liam Corcoran, "Here's How Newsrooms Should Be Rethinking Social Metrics," *NewsWhip*, May 31, 2017, https://www.newswhip.com/2017/05/newsrooms -thinking-differently-social-metrics/.

10. Adam Smith, "A Close Friend, or a Critical One?," Medium, October 25, 2017, https://medium.com/severe-contest/a-close-friend-or-a-critical-one-f4d1cc058114.

11. For further discussion, see, for example, Julia Greenberg, "Buzzfeed Is Changing the Way It Measures Its Popularity," *Wired*, February 18, 2016, https://www.wired .com/2016/02/buzzfeed-is-changing-the-way-it-measures-its-popularity/.

12. See, for example, Richard J. Tofel, "Non-profit Journalism: Issues around Impact," ProPublica, accessed May 1, 2018, https://s3.amazonaws.com/propublica/ assets/about/LFA_ProPublica-white-paper_2.1.pdf.

13. Alex Pentland, *Social Physics: How Good Ideas Spread—the Lessons from a New Science* (New York: Penguin Books, 2014).

14. Rasmus Kleis Nielsen and Federica Cherubini, "Editorial Analytics: How News Media Are Developing and Using Audience Data and Metrics," Reuters Institute for the Study of Journalism, University of Oxford, 2016, https://reutersinstitute.politics .ox.ac.uk/our-research/editorial-analytics-how-news-media-are-developing-and -using-audience-data-and-metrics.

15. See a list of research at the University of Texas at Austin Center for Media Engagement at https://mediaengagement.org/research/; and Wihbey, "Challenges of Democratizing News and Information."

16. Alexis Sobel Fitts, "Can Tony Haile Save Journalism by Changing the Metric?," *Columbia Journalism Review*, March 11, 2015, https://www.cjr.org/innovations/ tony_haile_chartbeat.php.

17. Daniel Mintz, "3 Interesting Things Attention Minutes Have Already Taught Us," *Upworthy Insider* (blog), February 13, 2014, https://blog.upworthy.com/post/ 76538569963/3-interesting-things-attention-minutes-have-already.

18. Tom Rosenstiel, "Solving Journalism's Hidden Problem: Terrible Analytics," Center for Effective Public Management at Brookings, February 2016, https://www .brookings.edu/wp-content/uploads/2016/07/Solving-journalisms-hidden-problem .pdf.

19. "Social, Search & Direct: Pathways to Digital News," Pew Research Center, March 2014, http://assets.pewresearch.org/wp-content/uploads/sites/13/2014/03/Social-Search -and-Direct-Pathways-to-Digital-News-copy-edited.pdf; and Tony Haile, "What You

Think You Know about the Web Is Wrong," *Time*, March 9, 2014, http://time.com/12933/what-you-think-you-know-about-the-web-is-wrong/.

20. Rodrigo Zamith, "Quantified Audiences in News Production," *Digital Journalism* 6, no. 4 (2018): 418–435, https://doi.org/10.1080/21670811.2018.1444999.

21. Matthew Hindman, "Journalism Ethics and Digital Audience Data," in *Remaking the News: Essays on the Future of Journalism Scholarship in the Digital Age*, ed. C. W. Anderson and Pablo J. Boczkowki (Cambridge, MA: MIT Press, 2017), 185–186.

9 News and Democracy

1. Moses, "'Never More Accurate.'"

2. Lee B. Becker and Mengtian Chen, "Public Trust in Journalism and Media: Analysis of Data from 1970 to 2015," Kettering Foundation Working Paper [2016:2], Charles F. Kettering Foundation, Dayton, OH, February 1, 2016, https://www.kettering.org/sites/default/files/product-downloads/Public%20Trust%20in%20Journalism%20and%20Media.pdf.

3. Justin McCarthy, "Trust in Mass Media Returns to All-Time Low," Gallup News, September 17, 2014, http://news.gallup.com/poll/176042/trust-mass-media-returns-time-low.aspx.

4. Becker and Chen, "Public Trust in Journalism and Media."

5. Jonathan M. Ladd, *Why Americans Hate the Media and How It Matters* (Princeton, NJ: Princeton University Press, 2012), 108–127; Becker and Chen, "Public Trust in Journalism and Media," 18.

6. Matthew A. Baum, *Soft News Goes to War: Public Opinion and American Foreign Policy in the New Media Age* (Princeton, NJ: Princeton University Press, 2005); Ladd, *Why Americans Hate the Media*, 110.

7. Michael Barthel and Amy Mitchell, "Americans' Attitudes about the News Media Deeply Divided along Partisan Lines," Pew Research Center, May 10, 2017, http://www.journalism.org/2017/05/10/americans-attitudes-about-the-news-media-deeply-divided-along-partisan-lines/#.

8. Schudson, *Power of News*, 171–172.

9. Spiro Agnew, speech at the Midwest Republican Conference, November 13, 1969, http://www.americanrhetoric.com/speeches/spiroagnewtvnewscoverage.htm.

10. Albert C. Gunther and Kathleen Schmitt, "Mapping Boundaries of the Hostile Media Effect," *Journal of Communication* 54, no. 1 (March 2004): 55–70, https://onlinelibrary.wiley.com/doi/10.1111/j.1460-2466.2004.tb02613.x/full.

11. Thomas E. Patterson, *Out of Order: An Incisive and Boldly Original Critique of the News Media's Domination of America's Political Process* (New York: First Vintage Books Edition, 1994).

12. Schudson, *Power of News*, 171–172.

13. Kevin Barnhurst, "Problem of Realist Events in American Journalism."

14. Patterson, "Doing Well and Doing Good."

15. Jackie Calmes, "'They Don't Give a Damn about Governing': Conservative Media's Influence on the Republican Party," Shorenstein Center on Media, Politics and Public Policy, Harvard University Discussion Series Paper D-96, July 2015.

16. Richard F. Fenno Jr., *Home Style: House Members in Their Districts* (Harlow, UK: Longman, 1978).

17. Ladd, *Why Americans Hate the Media*, 7.

18. James E. Campbell, *Polarized: Making Sense of a Divided America* (Princeton, NJ: Princeton University Press, 2016).

19. Natalie Jomini Stroud, "Polarization and Partisan Selective Exposure," *Journal of Communication* 60 (2010), 556–576, https://doi.org/10.1111/j.1460-2466.2010.01497.x.

20. Alan I. Abramowitz and Kyle L. Saunders, "Is Polarization a Myth?," *The Journal of Politics* 70, no. 2 (April 2008), 542–555, https://doi.org/10.1017/S0022381608080493.

21. Gentzkow and Shapiro, "What Drives Media Slant?"

22. W. Phillips Davison, "The Third-Person Effect in Communication," *Public Opinion Quarterly* 47, no. 1 (1983): 1–15.

23. Jean M. Twenge, Keith W. Campbell, and Nathan T. Carter, "Declines in Trust in Others and Confidence in Institutions among American Adults and Late Adolescents, 1972–2012," *Psychological Science* 25, no. 10 (2014): 1914–1923, https://doi.org/10.1177/0956797614545133.

24. Lelkes, "Mass Polarization"; and Douglas J. Ahler, "Self-Fulfilling Misperceptions of Public Polarization," *Journal of Politics* 76, no. 3 (2014): 607–620, https://doi.org/10.1017/S0022381614000085.

25. "The Personal News Cycle," Media Insight Project, American Press Institute and the Associated Press-NORC Center for Public Affairs Research, March 17, 2014.

26. Natalie Jomini Stroud, *Niche News: The Politics of News Choice* (New York: Oxford University Press, 2011), 176.

27. Diana C. Mutz, *Hearing the Other Side: Deliberative versus Participatory Democracy* (New York: Cambridge University Press, 2006), 16.

28. Kevin Arceneaux and Martin Johnson, *Changing Minds or Changing Channels? Partisan News in an Age of Choice* (Chicago: Chicago University Press, 2013), 88.

29. Cass Sunstein, *Going to Extremes: How Like Minds Unite and Divide* (New York: Oxford University Press, 2009).

30. Arceneaux and Johnson, *Changing Minds or Changing Channels?*, 88.

31. Markus Prior, *Post-Broadcast Democracy: How Media Choice Increases Inequality in Political Involvement and Polarizes Elections* (New York: Cambridge University Press, 2007), 256.

32. Yochai Benkler et al., "Study: Breitbart-Led Right-Wing Media Ecosystem Altered Broader Media Agenda," *Columbia Journalism Review*, March 3, 2017, https://www.cjr.org/analysis/breitbart-media-trump-harvard-study.php.

33. The Center for Information & Research on Civic Learning and Engagement, "An Estimated 24 Million Young People Voted in 2016 Election," Tufts University,

November 9, 2016, https://civicyouth.org/an-estimated-24-million-young-people-vote-in-2016-election/.

34. Levi Boxell, Matthew Gentzkow, and Jesse M. Shapiro, "Greater Internet Use Is Not Associated with Faster Growth in Political Polarization among US Demographic Groups," *Proceedings of the National Academy of Sciences of the United States of America* 114, no. 40 (October 3, 2017): 10612–10617, https://doi.org/10.1073/pnas.170658 8114.

35. Gregory J. Martin and Ali Yurukoglu, "Bias in Cable News: Persuasion and Polarization," *American Economic Review* 107, no. 9 (September 2017): 2565–2599, https://doi.org/10.1257/aer.20160812.

36. Andrew Kohut, Robert C. Toth, and Carol Bowman, "Public Tunes Out Recent News," Times Mirror Center for the People & the Press, May 19, 1994.

37. "Younger Americans and Women Less Informed: One in Four Americans Follow National News Closely," Pew Research Center, December 28, 1995.

38. William A. Galston, "Political Knowledge, Political Engagement and Civic Education," *Annual Review of Political Science* 4 (June 2001): 217–234, https://doi.org/10.1146/annurev.polisci.4.1.217.

39. "What the Public Knows—In Pictures, Words, Maps and Graphs," Pew Research Center, April 28, 2015.

40. Prior, *Post-Broadcast Democracy*, 94–137.

41. Scott L. Althaus, "American News Consumption during Times of National Crisis," *PS: Political Science & Politics* 35, no. 3 (September 2002), 517–521, http://www.jstor.org/stable/1554680.

42. Claire Wardle, "6 Types of Misinformation Circulated This Election Season," *Columbia Journalism Review*, November 18, 2016, https://www.cjr.org/tow_center/6_types_election_fake_news.php?CJR.

43. Allcott and Gentzkow, "Social Media and Fake News."

44. Christopher Paul and Miriam Matthews, "The Russian 'Firehose of Falsehood' Propaganda Model," RAND Corporation, Perspectives, 2016, https://www.rand.org/content/dam/rand/pubs/perspectives/PE100/PE198/RAND_PE198.pdf.

45. Panagiotis Takis Metaxas, "Web Spam, Social Propaganda and the Evolution of Search Engine Rankings," 2009, http://cs.wellesley.edu/~pmetaxas/Metaxas-EvolutionSEs-LNBIP10.pdf.

46. Amy Mitchell et al., "How Americans Encounter, Recall and Act upon Digital News," Pew Research Center, February 9, 2017, http://www.journalism.org/2017/02/09/how-americans-encounter-recall-and-act-upon-digital-news/.

47. Lee Shaker, "Dead Newspapers and Citizens' Civic Engagement," *Political Communication* 31, no. 1 (2014): 131–148, https://doi.org/10.1080/10584609.2012.762817.

48. Sam Schulhofer-Wohl and Miguel Garrido, "Do Newspapers Matter? Evidence from the Closure of *The Cincinnati Post*," Princeton University, Woodrow Wilson School of Public and International Affairs, Discussion Papers in Economics, No. 236, October 2009.

49. Waldman and the Working Group on Information Needs of Communities Federal Communications Commission, *Information Needs of Communities*.

50. Danny Hayes and Jennifer L. Lawless, "As Local News Goes, So Goes Citizen Engagement: Media, Knowledge, and Participation in US House Elections," *Journal of Politics* 77, no. 2 (April 2015): 447–462, https://doi.org/10.1086/679749.

51. Penelope Muse Abernathy, "The Rise of a New Media Baron and the Emerging Threat of News Deserts," Center for Innovation & Sustainability in Local Media, University of North Carolina School of Media and Journalism, 2016, http://newspaperownership.com/.

52. Joshua Benton, "As Giant Platforms Rise, Local News Is Getting Crushed," Nieman Lab, September 1, 2015, www.niemanlab.org/2015/09/as-giant-platforms -rise-local-news-is-getting-crushed/.

53. Steve Coll, "A Hole in the Heart of American Journalism," *Columbia Journalism Review*, Spring 2016, https://www.cjr.org/analysis/a_hole_in_the_heart_of_american _journalism.php.

Conclusion

1. Hardin, *How Do You Know?*

2. J. M. Coetzee and Arabella Kurtz, *The Good Story* (New York: Penguin Books, 2015), 105.

3. David Riesman, with Nathan Glazer and Reuel Denney, *The Lonely Crowd: A Study of the Changing American Character*, rev. ed. (New Haven, CT: Yale University Press, 2001), 99.

4. Christopher Mele, "Fatigued by the News? Experts Suggest How to Adjust Your Media Diet," *New York Times*, February 1, 2017, https://www.nytimes.com/2017/02/ 01/us/news-media-social-media-information-overload.html.

5. Michael Barthel, Amy Mitchell, and Jesse Holcomb, "Many Americans Believe Fake News Is Sowing Confusion," Pew Research Center, December 15, 2016, http://www .journalism.org/2016/12/15/many-americans-believe-fake-news-is-sowing-confusion/; and Ahler, "Self-Fulfilling Misperceptions of Public Polarization."

6. William Faulkner, "William Faulkner—Banquet Speech," speech at the Nobel Banquet at the City Hall in Stockholm, December 10, 1950, https://www.nobelprize .org/nobel_prizes/literature/laureates/1949/faulkner-speech.html.

7. See, for example, Mike Ferguson, "What Is Graph Analytics? Making the Complex Simple," IBM Big Data & Analytics Hub, May 17, 2016, http://www.ibmbigdatahub .com/blog/what-graph-analytics; Mike Hoskins, "How Graph Analytics Deliver Deeper Understanding," *InfoWorld*, February 4, 2015, https://www.infoworld.com/ article/2877489/big-data/how-graph-analytics-delivers-deeper-understanding-of -complex-data.html; and Emil Eifrem, "Graph Theory: Key to Understanding Big Data," *Wired*, May 2014, https://www.wired.com/insights/2014/05/graph-theory- key-understanding-big-data-2/.

8. Douglas Clement, "Interview with Matthew Gentzkow," *Region*, Federal Reserve Bank of Minneapolis, May 23, 2016, https://www.minneapolisfed.org/publications/the-region/interview-with-matthew-gentzkow.

9. Associated Press v. U.S. 57 S. Ct. (1945), https://caselaw.findlaw.com/us-supreme-court/326/1.html.

10. Starr, *Creation of the Media*.

11. Tim Wu, *The Master Switch: The Rise and Fall of Information Empires* (New York: Vintage Books, 2011).

12. Ladd, *Why Americans Hate the Media*, 8.

13. Victor Pickard, "Rediscovering the News: Journalism Studies' Three Blind Spots," in *Remaking the News: Essays on the Future of Journalism Scholarship in the Digital Age*, ed. C. W. Anderson and Pablo J. Boczkowki (Cambridge, MA: MIT Press, 2017), 52–60.

14. Starr, *Creation of the Media*.

15. James Madison, "For the National Gazette, [ca. 19 December] 1791," Founders Online, National Archives, last modified June 29, 2017, https://founders.archives.gov/documents/Madison/01-14-02-0145; emphasis in original.

16. Nicholas Lemann, "Solving the Problem of Fake News," *New Yorker*, November 30, 2016, https://www.newyorker.com/news/news-desk/solving-the-problem-of-fake-news.

17. Sunstein, *#Republic*, 211.

18. Adam Hersh, "Slowing Down the Presses: The Relationship between Net Neutrality and Local News," Center for Internet and Society, Stanford University, December 12, 2017, https://cyberlaw.stanford.edu/files/publication/files/20171208-NetNeutralityandLocalNews.pdf.

19. Matthew Nisbet, John Wihbey, Silje Kristiansen, and Aleszu Bajak, "Funding the News: Foundations and Nonprofit Media," Shorenstein Center on Media, Politics and Public Policy, June 18, 2018, https://shorensteincenter.org/funding-the-news-foundations-and-nonprofit-media/.

20. Leonard Downie Jr. and Michael Schudson, "The Reconstruction of American Journalism," *Columbia Journalism Review*, November/December 2009, https://archives.cjr.org/reconstruction/the_reconstruction_of_american.php.

21. Dan Kennedy, "Beyond the Tea Party: How the IRS Is Killing Nonprofit Media," *HuffPost*, July 16, 2013, https://www.huffingtonpost.com/dan-kennedy/beyond-the-tea-party-how_b_3286604.html.

22. Huey, Nisenholtz, and Sagan, "Riptide."

Bibliography

Abernathy, Penelope Muse. "The Rise of a New Media Baron and the Emerging Threat of News Deserts." Center for Innovation & Sustainability in Local Media, University of North Carolina School of Media and Journalism, 2016. http://news paperownership.com/.

Abramowitz, Alan I., and Kyle L. Saunders. "Is Polarization a Myth?" *Journal of Politics* 70, no. 2 (April 2008): 542–555.

Achen, Christopher H., and Larry M. Bartels. *Democracy for Realists: Why Elections Do Not Produce Responsive Government.* Princeton, NJ: Princeton University Press, 2016.

Adam, G. Stuart. "The Education of Journalists." *Journalism* 2, no. 3 (December 1, 2001): 315–339. https://doi.org/10.1177/146488490100200309.

Adamic, L. A., and E. Adar. "Friends and Neighbors on the Web." *Social Networks* 25, no. 3 (2003): 211–230. https://doi.org/10.1016/S0378-8733(03)00009-1.

Adams, Tim. "And the Pulitzer Goes to ... a Computer." *Guardian*, June 28, 2015. https://www.theguardian.com/technology/2015/jun/28/computer-writing-journal ism-artificial-intelligence.

Advisory Committee on Public Interest Obligations of Digital Television Broadcasters. *Charting the Digital Broadcasting Future: Final Report of the Advisory Committee on Public Interest Obligations of Digital Television Broadcasters.* Washington, DC: The Benton Foundation, 1998. http://govinfo.library.unt.edu/piac/piacreport.pdf.

Agnew, Spiro. Speech at the Midwest Republican Conference, November 13, 1969. http://www.americanrhetoric.com/speeches/spiroagnewtvnewscoverage.htm.

Ahler, Douglas J. "Self-Fulfilling Misperceptions of Public Polarization." *Journal of Politics* 76, no. 3 (2014): 607–620. https://doi.org/10.1017/S0022381614000085.

Al-Rawi, Ahmed. "Viral News on Social Media." *Digital Journalism*, October 2017, 1–17. https://doi.org/10.1080/21670811.2017.1387062.

Alberta, Tim. "The Deep Roots of Trump's War on the Press." *Politico*, April 26, 2018. https://www.politico.com/magazine/story/2018/04/26/the-deep-roots-trumps-war -on-the-press-218105.

Alexander, Jeffrey C. "Journalism, Democratic Culture, and Creative Reconstruction." In *The Crisis of Journalism Reconsidered: Democratic Culture, Professional Codes, Digital Futures*, ed. Jeffrey C. Alexander, Elizabeth Butler Breese, and Maria Luengo, 1–28. Cambridge: Cambridge University Press, 2016.

Allcott, Hunt, and Matthew Gentzkow. "Social Media and Fake News in the 2016 Election." *Journal of Economic Perspectives* 31, no. 2 (Spring 2017): 211–236. https:// doi.org/10.1257/jep.31.2.211.

Allen, Danielle. *Talking to Strangers: Anxieties of Citizenship since Brown v. Board of Education.* Chicago: University of Chicago Press, 2004.

Allocca, Kevin. *Videocracy: How YouTube Is Changing the World … with Double Rainbows, Singing Foxes, and Other Trends We Can't Stop Watching.* London: Bloomsbury, 2018.

Althaus, Scott L. "American News Consumption during Times of National Crisis." *PS: Political Science & Politics* 35, no. 3 (September 2002): 517–521. http://www.jstor .org/stable/1554680.

Amazeen, Michelle A., Emily Thorson, Ashley Muddiman, and Lucas Graves. "Correcting Political and Consumer Misperceptions: The Effectiveness and Effects of Rating Scale versus Contextual Correction Formats." *Journalism & Mass Communications Quarterly* 95, no. 1 (Spring 2018): 28–48. https://doi.org/10.1177/10776990166 78186.

Ananny, Mike. "Networked Press Freedom and Social Media: Tracing Historical and Contemporary Forces in Press-Public Relations." *Journal of Computer-Mediated Communication* 19, no. 4 (July 2014): 938–956. https://doi.org/10.1111/jcc4.12076.

Ananny, Mike. *Networked Press Freedom: Creating Infrastructures for a Public Right to Hear.* Cambridge, MA: MIT Press, 2018.

Anderson, Ashley A., Dominique Brossard, Dietram A. Scheufele, Michael A. Xenos, and Peter Ladwig. "The 'Nasty Effect': Online Incivility and Risk Perceptions of Emerging Technologies." *Journal of Computer-Mediated Communication* 19, no. 3 (February 19, 2013): 373–387. https://doi.org/10.1111/jcc4.12009.

Anderson, C. W. *Rebuilding the News: Metropolitan Journalism in the Digital Age.* Philadelphia: Temple University Press, 2013.

Angwin, Julia, Jeff Larson, Surya Mattu, and Lauren Kirchner. "Machine Bias." ProPublica, May 23, 2016. https://www.propublica.org/article/machine-bias-risk-assess ments-in-criminal-sentencing.

Arceneaux, Kevin, and Martin Johnson. *Changing Minds or Changing Channels? Partisan News in an Age of Choice.* Chicago: Chicago University Press, 2013.

Ariens, Chris. "Evening News Ratings: Week of January 16." *AdWeek*, January 24, 2017.

"Artificial Intelligence: Practice and Implications for Journalism." Conference event streamed by Columbia Journalism School on June 13, 2017. YouTube video. https://www.youtube.com/watch?v=vwaJKJ-zya0.

Asch, Solomon E. "Studies of Independence and Conformity: A Minority of One against a Unanimous Majority." *Psychological Monographs: General and Applied* 70, no. 9 (1956): 1–70. http://psycnet.apa.org/record/2011-16966-001.

Associated Press-NORC Center for Public Affairs Research. "Confidence in Institutions: Trends in Americans' Attitudes toward Government, Media, and Business." Associated Press and NORC, March 2015. http://www.apnorc.org/projects/Pages/confidence-in-institutions-trends-in-americans-attitudes-toward-government-media-and-business.aspx.

Associated Press v. U.S. 57 S. Ct (1945). https://caselaw.findlaw.com/us-supreme-court/326/1.html.

Atlanta Journal-Constitution. "How the Doctors and Sex Abuse Project Came About." *Atlanta Journal-Constitution*, 2016. http://doctors.ajc.com/about_this_investigation/.

Austin, Anne, Jonathan Barnard, and Nicola Hutcheon. "Media Consumption Forecasts 2016." Zenith, the ROI Agency, June 2016. https://communicateonline.me/wp-content/uploads/2016/06/Media-Consumption-Forecasts-2016.pdf.

Backstrom, Lars, Paolo Boldi, Marco Rosa, Johan Ugander, and Sebastiano Vigna. "Four Degrees of Separation." Paper presented at the Third Annual ACM Web Science Conference, Evanston, Illinois, June 22–24, 2012. https://doi.org/10.1145/2380718.2380723.

Badger, Emily, and Quoctrung Bui. "In 83 Million Eviction Records, a Sweeping and Intimate New Look at Housing in America." *New York Times*, April 7, 2018. https://www.nytimes.com/interactive/2018/04/07/upshot/millions-of-eviction-records-a-sweeping-new-look-at-housing-in-america.html.

Bajak, Aleszu. "Should Newspapers Be Adding Confidence Intervals to Their Graphics?" Storybench, May 19, 2017. http://www.storybench.org/newspapers-adding-confidence-intervals-graphics/.

Bajak, Aleszu. "Topic Modeling the 10,000 Top-Performing News Articles on Social Media Published between Nov. 2016 and May 2017." Github. https://github.com/aleszu/topicmodeling/blob/master/README.md.

Bajak, Aleszu. "Using BuzzFeed's Listicle Format to Tell Stories with Maps and Charts." Storybench, August 7, 2015. http://www.storybench.org/using-buzzfeeds -listicle-format-tell-stories-maps-charts/.

Baker, Peter, and Celia Kang. "Trump Threatens NBC Over Nuclear Weapons Report." *New York Times*, October 11, 2017. https://www.nytimes.com/2017/10/11/ us/politics/trump-nbc-fcc-broadcast-license.html.

Bakshy, Eytan, Dean Eckles, Rong Yan, and Itamar Rosenn. "Social Influence in Social Advertising: Evidence from Field Experiments." In *Proceedings of the 13th ACM Conference on Electronic Commerce*, 146–161. New York: ACM, 2012. https://doi .org/10.1145/2229012.2229027.

Bakshy, Eytan, Solomon Messing, and Lada Adamic. "Exposure to Ideologically Diverse News and Opinion on Facebook." *Science* 348, no. 6239 (June 2015): 1130– 1132. https://doi.org/10.1126/science.aaa1160.

Bakshy, Eytan, Itamar Rosenn, Cameron Marlow, and Lada Adamic. "The Role of Social Networks in Information Diffusion." In *Proceedings of the 21st International Conference on World Wide Web*, 519–528. New York: ACM, 2012. https://doi.org/ 10.1145/2187836.2187907.

Barabási, Albert-László. *Bursts: The Hidden Patterns Behind Everything We Do, from Your E-mail to Bloody Crusades*. New York: Dutton, 2010.

Barabási, Albert-László. *Linked: The New Science of Networks*. Cambridge, MA: Perseus Books Group, 2002.

Barabási, Albert-László. *Network Science*. Cambridge: Cambridge University Press, 2016. Available under a Creative Commons License at http://barabasi.com/network sciencebook/.

Barabási, Albert-László. "Scale-Free Networks: A Decade and Beyond." *Science* 325, no. 5939 (July 2009): 412–413. https://doi.org/10.1126/science.1173299.

Barberá, Pablo, John T. Jost, Jonathan Nagler, Joshua A. Tucker, and Richard Bon- neau. "Tweeting from Left to Right: Is Online Political Communication More Than an Echo Chamber?" *Psychological Science* 26, no. 10 (August 2015): 1–12. https://doi .org/10.1177/0956797615594620.

Barnhurst, Kevin G. "The Problem of Realist Events in American Journalism." *Media and Communication* 2, no. 2 (October 2014): 84–95. https://doi.org/10.17645/mac .v2i2.159.

Barnhurst, Kevin G., and Diana Mutz. "American Journalism and the Decline in Event-Centered Reporting." *Journal of Communication* 47, no. 4 (1997): 27–53.

Barr, Jeremy. "The *New York Times* Pulls Back Ahead of the *Washington Post* for Unique Visitors." *Advertising Age*, February 17, 2017. http://adage.com/article/media/ york-times-pulls-back-ahead-washington-post/302720/.

Barthel, Michael, and Amy Mitchell. "Americans' Attitudes about the News Media Deeply Divided along Partisan Lines." Pew Research Center, May 10, 2017. http://www.journalism.org/2017/05/10/americans-attitudes-about-the-news-media-deeply-divided-along-partisan-lines/#.

Barthel, Michael, Amy Mitchell, and Jesse Holcomb. "Many Americans Believe Fake News Is Sowing Confusion." Pew Research Center, December 15, 2016. http://www.journalism.org/2016/12/15/many-americans-believe-fake-news-is-sowing-confusion/.

Batsell, Jake. *Engaged Journalism: Connecting with Digitally Empowered News Audiences.* New York: Columbia University Press, 2015.

Battelle, John. *The Search: How Google and Its Rivals Rewrote the Rules of Business and Transformed Our Culture.* New York: Penguin, 2005.

Baum, Matthew A. *Soft News Goes to War: Public Opinion and American Foreign Policy in the New Media Age.* Princeton, NJ: Princeton University Press, 2005.

Baum, Matthew A., and David Lazer. "Google and Facebook Aren't Fighting Fake News with the Right Weapons." *Los Angeles Times*, May 8, 2017. http://www.latimes.com/opinion/op-ed/la-oe-baum-lazer-how-to-fight-fake-news-20170508-story.html.

Beaudet, Mike, John Wihbey, Matt Tota, Aneri Pattani, Sophia Fox-Sowell, Emily Hopkins, Emily Turner, Gail Waterhouse, Irene de la Torre-Arenas, and Pedro Cruz. "The State Financial Disclosure Project." Northeastern University College of Arts, Media and Design, 2016. https://web.northeastern.edu/disclosure-project/.

Becker, Lee B., and Mengtian Chen. "Public Trust in Journalism and Media: Analysis of Data from 1970 to 2015." Kettering Foundation Working Paper [2016:2], Charles F. Kettering Foundation, Dayton, Ohio, February 1, 2016. https://www.kettering.org/sites/default/files/product-downloads/Public%20Trust%20in%20Journalism%20and%20Media.pdf.

Bell, Daniel. *The Coming of Post-Industrial Society: A Venture in Social Forecasting.* New York: Basic Books, 1973.

Bell, Emily. "Who Owns the News Consumer: Social Media Platforms or Publishers?" *Columbia Journalism Review*, June 21, 2016. https://www.cjr.org/tow_center/platforms_and_publishers_new_research_from_the_tow_center.php.

Bell, Emily, C. W. Anderson, and Clay Shirky. "Post Industrial Journalism: Adapting to the Present." Tow Center for Digital Journalism, New York, December 3, 2014. https://doi.org/10.7916/D8N01JS7.

Benkler, Yochai. *The Wealth of Networks: How Social Production Transforms Markets and Freedom.* New Haven, CT: Yale University Press, 2006.

Benkler, Yochai, Robert Faris, Hal Roberts, and Ethan Zuckerman. "Study: Breitbart-Led Right-Wing Media Ecosystem Altered Broader Media Agenda." *Columbia Journalism Review*, March 3, 2017. https://www.cjr.org/analysis/breitbart-media-trump-harvard-study.php.

Bennett, W. Lance, and Alexandra Segerberg. "The Logic of Connective Action." *Information, Communication and Society* 15, no. 5 (April 10, 2012): 739–768. http://ccce.com.washington.edu/about/assets/2012iCS-LCA-Bennett&Segerberg-Logicof ConnectiveAction.pdf.

Bennett, W. Lance, and Alexandra Segerberg. *The Logic of Connective Action: Digital Media and the Personalization of Contentious Politics*. Cambridge: Cambridge University Press, 2013.

Benton, Joshua. "As Giant Platforms Rise, Local News is Getting Crushed." Nieman Lab, September 1, 2015. www.niemanlab.org/2015/09/as-giant-platforms-rise-local-news-is-getting-crushed/.

Bertini, Enrico, and Moritz Stefaner. "Uncertainty and Trumpery with Alberto Cairo." *Data Stories*, March 21, 2017. Podcast audio. http://datastori.es/94-uncertainty-and-trumpery-with-alberto-cairo/.

Bever, Lindsey. "Memo to Donald Trump: Thomas Jefferson Invented Hating the Media." *Washington Post*, February 18, 2017. https://www.washingtonpost.com/news/the-fix/wp/2017/02/17/trumps-war-with-the-media-isnt-new-thomas-jefferson-railed-about-newspaper-lies-too/.

Bhagat, Smriti, Moira Burke, Carlos Diuk, Ismail Onur Filiz, and Sergey Edunov. "Three and a Half Degrees of Separation." Facebook Research (blog), February 4, 2016. https://research.fb.com/three-and-a-half-degrees-of-separation/.

Bilton, Ricardo. "Reuters Built Its Own Algorithmic Prediction Tool to Help It Spot (and Verify) Breaking News on Twitter." Nieman Lab, November 30, 2016. http://www.niemanlab.org/2016/11/reuters-built-its-own-algorithmic-prediction-tool-to-help-it-spot-and-verify-breaking-news-on-twitter/.

Boczkowski, Pablo J., and Eugenia Mitchelstein. "Scholarship on Online Journalism: Roads Traveled and Pathways Ahead." In *Remaking the News: Essays on the Future of Journalism Scholarship in the Digital Age*. ed. C. W. Anderson and Pablo J. Boczkowki, 15–26. Cambridge, MA: MIT Press, 2017.

Boczkowski, Pablo J., and Eugenia Mitchelstein. *The News Gap: When the Information Preferences of the Media and the Public Diverge*. Cambridge, MA: MIT Press, 2015.

Bond, Robert M., Christopher J. Fariss, Jason J. Jones, Adam D. I. Kramer, Cameron Marlow, Jaime E. Settle, and James H. Fowler. "A 61-Million-Person Experiment in Social Influence and Political Mobilization." *Nature* 489, no. 7415 (September 2012): 295–298. https://doi.org/10.1038/nature11421.

Borgatti, Stephen P., Ajay Mehra, Daniel J. Brass, and Giuseppe Labianca. "Network Analysis in the Social Sciences." *Science* 323, no. 892 (February 13, 2009): 892–895. https://doi.org/10.1126/science.1165821.

Bostock, Michael, Vadim Ogievetsky, and Jeffrey Heer. "D3: Data-Driven Documents." *IEEE Transactions on Visualization & Computer Graphics Proc. InfoVis* (October 2011). http://vis.stanford.edu/papers/d3.

Bounegru, Liliana, Tommaso Venturini, Jonathan Gray, and Mathieu Jacomy. "Narrating Networks." *Digital Journalism* 5, no. 6 (2017) 699–730. https://doi.org/10.1080/21670811.2016.1186497.

Boxell, Levi, Matthew Gentzkow, and Jesse M. Shapiro. "Greater Internet Use Is Not Associated with Faster Growth in Political Polarization among US Demographic Groups." *Proceedings of the National Academy of Sciences of the United States of America* 114, no. 40 (October 3, 2017): 10612–10617. https://doi.org/10.1073/pnas.1706588114.

Boxell, Levi, Matthew Gentzkow, and Jesse M. Shapiro. "Is the Internet Causing Political Polarization? Evidence from Demographics." NBER Working Paper no. 23258, March 2017. https://www.brown.edu/Research/Shapiro/pdfs/age-polars.pdf.

Boykoff, Maxwell T., and Jules M. Boykoff. "Balance as Bias: Global Warming and the US Prestige Press." *Global Environmental Change* 14, no. 2 (July 2004): 125–136. https://doi.org/10.1016/j.gloenvcha.2003.10.001.

Boyte, Harry C. *Civic Agency and the Cult of the Expert*. Dayton, OH: Charles F. Kettering Foundation, 2009. https://files.eric.ed.gov/fulltext/ED510128.pdf.

Brandeis, Louis D. *Other People's Money, and How the Bankers Use It*. New York: Frederick A. Stokes, 1914.

Broussard, Meredith. *Artificial Unintelligence: How Computers Misunderstand the World*. Cambridge, MA: MIT Press, 2018.

Bruner, Jerome. "The Narrative Construction of Reality." *Critical Inquiry* 18, no. 1 (Autumn 1991): 1–21. https://doi.org/10.1086/448619.

Burke, Moira, and Robert Kraut. "Growing Closer on Facebook: Changes in Tie Strength through Social Network Site Use." In *Proceedings of the SIGCHI Conference on Human Factors in Computing Systems*, 4187–4196. New York: ACM, 2014. https://doi.org/10.1145/2556288.2557094.

Burrell, Ian. "BuzzFeed's Jonah Peretti: News Publishers Only Have Themselves to Blame for Losing Out to Google and Facebook." *Drum*, June 29, 2017. http://www.thedrum.com/opinion/2017/06/29/buzzfeeds-jonah-peretti-news-publishers-only-have-themselves-blame-losing-out.

Burt, Ronald S. "Structural Holes and Good Ideas." *American Journal of Sociology* 110, no. 2 (September 2004): 349–399. https://doi.org/10.1086/421787.

Bush, George W. "The Spirit of Liberty: At Home, In the World." Remarks given at the George W. Bush Institute, New York, October 19, 2017. Video recording. https://www.bushcenter.org/exhibits-and-events/events/2017/10/p-spirit-of-liberty.

Cabra, Mar. "How the ICIJ Used Neo4j to Unravel the Panama Papers." *Neo4j Blog*, May 12, 2016. https://neo4j.com/blog/icij-neo4j-unravel-panama-papers/.

Calmes, Jackie. "'They Don't Give a Damn about Governing': Conservative Media's Influence on the Republican Party." Shorenstein Center on Media, Politics and Public Policy, Harvard University Discussion Series Paper D-96, July 2015.

Campbell, James E. *Polarized: Making Sense of a Divided America*. Princeton, NJ: Princeton University Press, 2016.

Caplan, Lincoln. "Should Facebook and Twitter Be Regulated Under the First Amendment?" *Wired*, October 11, 2017. https://www.wired.com/story/should-facebook-and-twitter-be-regulated-under-the-first-amendment/.

Carey, James W. "A Cultural Approach to Communication." In *Communication as Culture: Essays on Media and Society*, 13–36. Boston: Unwin Hyman, 1989. http://web.mit.edu/21l.432/www/readings/Carey_CulturalApproachCommunication.pdf.

Carey, James W. "A Short History of Journalism for Journalists: A Proposal and Essay." *Harvard International Journal of Press/Politics* 12, no. 3 (January 1, 2007): 3–16. https://doi.org/10.1177/1081180X06297603.

Carey, James W. "Technology and Ideology: The Case of the Telegraph." In *Communication as Culture: Essays on Media and Society*, rev. ed., 155–177. New York: Routledge, 2009.

Carpenter, Serena. "An Application of the Theory of Expertise: Teaching Broad and Skill Knowledge Areas to Prepare Journalists for Change." *Journalism & Mass Communication Educator* 64 (September 1, 2009): 287–304. https://doi.org/10.1177/107769580906400305.

Carr, Nicholas. "Is Google Making Us Stupid? What the Internet Is Doing to Our Brains." *Atlantic*, July/August 2008. https://www.theatlantic.com/magazine/archive/2008/07/is-google-making-us-stupid/306868/.

Carrington, Damian. "BBC Apologizes over Interview with Climate Denier Lord Lawson." *Guardian*, October 24, 2017. https://www.theguardian.com/environment/2017/oct/24/bbc-apologises-over-interview-climate-sceptic-lord-nigel-lawson.

Carroll, Matt, Sacha Pfeiffer, and Michael Rezendes. "Cchhur Allowed Abuse by Priest for Years." *Boston Globe*, January 2, 2002. https://www.bostonglobe.com/

news/special-reports/2002/01/06/church-allowed-abuse-priest-for-years/cSHfGkTIrA
T25qKGvBuDNM/story.html.

Castells, Manuel. *Communication Power*. Oxford: Oxford University Press, 2009.

Center for Media Engagement. "Our Research." Moody College of Communication,
University of Texas at Austin. Accessed June 18, 2018. https://mediaengagement
.org/research/.

Center for Information & Research on Civic Learning and Engagement. "An Estimated 24 Million Young People Voted in 2016 Election." Center for Information &
Research on Civic Learning and Engagement, Tufts University, November 9, 2016.

Chadwick, Andrew. *The Hybrid Media System: Politics and Power*. Oxford: Oxford University Press, 2013.

Cheng, Justin. "Do Cascades Recur?" Paper presented at the International World
Wide Web Conference, Montreal, Canada, April 14, 2016. https://arxiv.org/pdf/
1602.01107.pdf.

Cheng, Justin, Cristian Danescu-Niculescu-Mizil, and Jure Leskovec. "How Community Feedback Shapes User Behavior." In *Proceedings of the Eighth International AAAI
Conference on Weblogs and Social Media*, 41–50. Palo Alto, CA: AAAI Press, 2014.
https://www.aaai.org/ocs/index.php/ICWSM/ICWSM14/paper/viewFile/8066/8104.

Chokkattu, Julien. "New Google Algorithm Lowers Search Rankings for Holocaust
Denial Sites." Digital Trends, December 25, 2016. https://www.digitaltrends.com/
web/google-search-holocaust/.

Christakis, Nicholas A., and James H. Fowler. "The Spread of Obesity in a Large
Social Network over 32 Years." *New England Journal of Medicine* 357 (July 26, 2007):
370–379. https://doi.org/10.1056/NEJMsa066082.

Clement, Douglas. "Interview with Matthew Gentzkow." *Region*, Federal Reserve
Bank of Minneapolis, May 23, 2016. https://www.minneapolisfed.org/publications/
the-region/interview-with-matthew-gentzkow.

Clifford, Stephanie. "Christian Science Paper to End Daily Print Edition." *New York
Times*, October 28, 2008. https://www.nytimes.com/2008/10/29/business/media/
29paper.html.

Clinton, Josh, and Carrie Roush. "Poll: Persistent Partisan Divide over 'Birther' Question." *NBCNews*, August 10, 2016. https://www.nbcnews.com/politics/2016-election/
poll-persistent-partisan-divide-over-birther-question-n627446.

Coddington, Mark. "Clarifying Journalism's Quantitative Turn." *Digital Journalism* 3,
no. 3 (2015): 331–348. https://doi.org/10.1080/21670811.2014.976400.

Coetzee, J. M., and Arabella Kurtz. *The Good Story*. New York: Penguin Books, 2015.

Cohn, Victor, Lewis Cope, and Deborah Cohn Runkle. *News and Numbers: A Writer's Guide to Statistics*. 3rd ed. Hoboken, NJ: Wiley-Blackwell, 2011.

Coll, Steve. "A Hole in the Heart of American Journalism." *Columbia Journalism Review*, Spring 2016. https://www.cjr.org/analysis/a_hole_in_the_heart_of_american _journalism.php.

Commission on Freedom of the Press. *A Free and Responsible Press*. Chicago: University of Chicago Press, 1947. https://archive.org/details/freeandresponsib029216mbp.

Cope, Bill, and Mary Kalantzis. "The Role of the Internet in Changing Knowledge Ecologies." *Arbor* 185, no. 737 (May–June 2009): 521–530. https://doi.org/10.3989/ arbor.2009.i737.309.

Corbett-Davies, Sam, Emma Pierson, Avi Feller, and Sharad Goel. "A Computer Program Used for Bail and Sentencing Decisions Was Labeled Biased against Blacks. It's Actually Not that Clear." *Washington Post*, October 17, 2016. https://www. washingtonpost.com/news/monkey-cage/wp/2016/10/17/can-an-algorithm-be -racist-our-analysis-is-more-cautious-than-propublicas/?utm_term=.89099c54847f.

Corcoran, Liam. "Get NewsWhip's Three Years of Social Data Report." NewsWhip, March 1, 2017. https://www.newswhip.com/2017/03/three-years-social-data-report/.

Corcoran, Liam. "Here's How Newsrooms Should Be Rethinking Social Metrics." NewsWhip, May 31, 2017. https://www.newswhip.com/2017/05/newsrooms -thinking-differently-social-metrics/.

Crutchfield, Robert D., and Gregory A. Weeks. "The Effects of Mass Incarceration on Communities of Color." *Issues in Science and Technology* 32, no. 1 (Fall 2015). http:// issues.org/32-1/the-effects-of-mass-incarceration-on-communities-of-color.

Cuillier, David. *Forecasting Freedom of Information: Why It Faces Problems—and How Experts Say They Could Be Solved*. Miami, FL: John S. and James L. Knight Foundation, 2017.

Daly, Christopher B. *Covering America: A Narrative History of a Nation's Journalism*. Amherst: University of Massachusetts Press, 2012.

Dance, Gabriel, and Tom Meagher. "Crime in Context." The Marshall Project, August 18, 2016. https://www.themarshallproject.org/2016/08/18/crime-in-context.

Das, Angelica. *Pathways to Engagement: Understanding How Newsrooms Are Working with Communities*. Public Square Program, Democracy Fund, May 2, 2017. https:// www.democracyfund.org/publications/pathways-to-engagement-understanding -how-newsrooms-are-working-with-communi.

Davis, Aaron C., and Wesley Lowery. "FBI Director Calls Lack of Data on Police Shootings 'Ridiculous,' 'Embarrassing.'" *Washington Post*, October 7, 2015. https:// www.washingtonpost.com/national/fbi-director-calls-lack-of-data-on-police

-shootings-ridiculous-embarrassing/2015/10/07/c0ebaf7a-6d16-11e5-b31c-d80d 62b53e28_story.html.

Davison, W. Phillips. "The Third-Person Effect in Communication." *Public Opinion Quarterly* 47, no. 1 (1983): 1–15.

de Tocqueville, Alexis. *Democracy in America.* Edited by J. P. Mayer. New York: Harper Perennial, 1966/1988.

Dean, Walter C., and Atiba Pertilla. "I-Teams and 'Eye Candy': The Reality of Local TV News." In *We Interrupt This Newscast: How to Improve Local News and Win Ratings, Too,* edited by Tom Rosenstiel, Marion Just, Todd L. Belt, Atiba Pertilla, Walter Dean, and Dante Chinni, 30–50. New York: Cambridge University Press, 2007. https://doi .org/10.1080/10584600801985755.

Dewey, John. *The Public and Its Problems.* New York: Henry Holt and Company, 1927. Reprint, Athens, OH: Swallow Press, 1954.

Dewey, John. "Public Opinion." *New Republic,* May 3, 1922.

Diakopoulos, Nicholas. "Algorithmic Accountability: Journalistic Investigation of Computational Power Structures." *Digital Journalism,* no. 3 (November 7, 2014): 398–415. https://doi.org/10.1080/21670811.2014.976411.

Diamond, Jeremy Scott. "How We Built an Interactive Graphic Using Carcinogen Data from the W.H.O." Storybench, November 12, 2015. http://www.storybench. org/how-we-built-an-interactive-graphic-using-carcinogen-data-from-the-w-h-o/.

Dixon, Travis L. "Good Guys Are Still Always in White? Positive Change and Continued Misrepresentation of Race and Crime on Local Television News." *Communication Research* 44, no. 6 (April 5, 2015): 775–792. https://doi.org/10.1177/00936502 15579223.

Doctor, Ken. "Newsonomics: The Halving of America's Daily Newsrooms." Neiman Lab, July 28, 2015. http://www.niemanlab.org/2015/07/newsonomics-the-halving -of-americas-daily-newsrooms/.

Doctor, Ken. "Newsonomics: Trump May Be the News Industry's Greatest Opportunity to Build a Sustainable Model." Nieman Lab, January 20, 2017. http://www .niemanlab.org/2017/01/newsonomics-trump-may-be-the-news-industrys -greatest-opportunity-to-build-a-sustainable-model/.

Dodds, Peter, Roby Muhamad, and Duncan Watts. "An Experimental Study of Search in Global Social Networks." *Science* 301, no. 5634 (2003): 827–829. https:// doi.org/10.1126/science.1081058.

Doig, Stephen K. "Reporting with the Tools of Social Science." *Nieman Reports,* March 15, 2008. https://niemanreports.org/articles/reporting-with-the-tools-of -social-science/.

Domingos, Pedro. *The Master Algorithm: How the Quest for the Ultimate Learning Machine Will Remake the World*. New York: Basic Books, 2015.

Donsbach, Wolfgang. "Journalism as the New Knowledge Profession and Consequences for Journalism Education." *Journalism* 15, no. 6 (July 2, 2013): 661–677. https://doi.org/10.1177/1464884913491347.

Downie, Leonard, Jr. "The Obama Administration and the Press." Committee to Protect Journalists, October 10, 2013. https://cpj.org/reports/2013/10/obama-and -the-press-us-leaks-surveillance-post-911.php.

Downie, Leonard, Jr., and Michael Schudson. "The Reconstruction of American Journalism." *Columbia Journalism Review*, November/December 2009. https:// archives.cjr.org/reconstruction/the_reconstruction_of_american.php.

Drummond, Caitlin, and Baruch Fischhoff. "Individuals with Greater Science Literacy and Education Have More Polarized Beliefs on Controversial Science Topics." *Proceedings of the National Academy of Sciences of the United States of America* 114, no. 36 (2017): 9587–9592. https://doi.org/10.1073/pnas.1704882114.

Dunbar, R. I. M., Valerio Arnaboldia, Marco Conti, and Andrea Passarella. "The Structure of Online Social Networks Mirrors Those in the Offline World." *Social Networks*, no. 43 (October 2015): 39–47. https://doi.org/10.1016/j.socnet.2015.04.005.

Dunkelman, Marc J. *The Vanishing Neighbor: The Transformation of American Community*. New York: W. W. Norton & Co, 2014.

Durkheim, Émile. *The Rules of Sociological Method*. Translated by W. D. Halls. New York: Free Press, 1982.

Edmonds, Rick. "News Media Alliance Seeks Antitrust Exemption to Negotiate a Better Deal with Facebook and Google." Poynter, July 10, 2017. https://www .poynter.org/2017/news-media-alliance-seeks-antitrust-exemption-to-negotiate -a-better-deal-with-facebook-and-google/466126/.

Edwards, Jim. "Planet Selfie: We're Now Posting a Staggering 1.8 Billion Photos Every Day." *Business Insider*, May 28, 2014. http://www.businessinsider.com/ were-now-posting-a-staggering-18-billion-photos-to-social-media-every-day-2014-5.

Eifrem, Emil. "Graph Theory: Key to Understanding Big Data." *Wired*, May 2014. https://www.wired.com/insights/2014/05/graph-theory-key-understanding-big -data-2/.

Elizabeth, Jane. "After a Decade, It's Time to Reinvent Social Media in Newsrooms." American Press Institute, November 14, 2017. https://www.americanpressinstitute .org/publications/reports/strategy-studies/reinventing-social-media/.

The Engaging News Project: Journalist Involvement in Comment Sections. University of Texas at Austin Center for Media Engagement, September 10, 2014. https:// www.democracyfund.org/media/uploaded/ENP_Comments_Report.pdf.

Entman, Robert M., and Andrew Rojecki. *The Black Image in the White Mind: Media and Race in America*. Chicago: University of Chicago Press, 2001.

Erdős, P., and A. Rényi. "On the Evolution of Random Graphs." *Mathematical Institute of the Hungarian Academy of Sciences*, no. 5 (1960): 17–61. https://pdfs.semantic scholar.org/4201/73e087bca0bdb31985e28ff69c60a129c8ef.pdf.

Errico, Marcus. "The Evolution of the Summary News Lead." *Media History Monographs* 1, no. 1 (1997). https://www.scripps.ohiou.edu/mediahistory/mhmjour1-1.htm.

Ettema, James S. "Journalism as Reason-Giving: Deliberative Democracy, Institutional Accountability, and the News Media's Mission." *Political Communication* 24, no. 2 (May 2007): 143–160. https://doi.org/10.1080/10584600701312860.

Executive Office of the President. "Try This at Home." White House Office of Science and Technology, January 2017. https://www.whitehouse.gov/sites/whitehouse .gov/files/images/Try%20This%20At%20Home-01-2017.pdf.

Facebook Data Science. "Can Cascades Be Predicted?" Facebook, February 28, 2014. https://www.facebook.com/notes/facebook-data-science/can-cascades-be-predicted/ 10152056491448859/.

Farrell, Henry. "New Problems, New Publics? Dewey and New Media." *Policy and Internet* 6, no. 2 (June 2014): 176–191. https://doi.org/10.1002/1944-2866.POI363.

Faulkner, William. "William Faulkner—Banquet Speech." Speech at the Nobel Banquet at the City Hall, Stockholm, December 10, 1950. https://www.nobelprize.org/ nobel_prizes/literature/laureates/1949/faulkner-speech.html.

Fenno, Richard F., Jr. *Home Style: House Members in Their Districts* Longman Classics Series, 1978/2003. Harlow, UK: Longman, 1978.

Ferguson, Mike. "What Is Graph Analytics? Making the Complex Simple." IBM Big Data & Analytics Hub, May 17, 2016. http://www.ibmbigdatahub.com/blog/what -graph-analytics.

Ferguson, Niall. *The Square and the Tower: Networks and Power, from the Freemasons to Facebook*. New York: Penguin Press, 2018.

Fink, Katherine, and Michael Schudson. "The Rise of Contextual Journalism, 1950s–2000s." *Journalism* 15, no. 1 (February 17, 2013): 3–20. https://doi.org/10 .1177/1464884913479015.

Fitts, Alexis Sobel. "Can Tony Haile Save Journalism by Changing the Metric?" *Columbia Journalism Review*, March 11, 2015. https://www.cjr.org/innovations/ tony_haile_chartbeat.php.

Flaxman, Seth, Sharad Goel, and Justin M. Rao. "Filter Bubbles, Echo Chambers, and Online News Consumption." *Public Opinion Quarterly* 80, no. S1 (2016): 298–320.

Floridi, Luciano. *The Fourth Revolution: How the Infosphere Is Reshaping Human Reality*. Oxford: Oxford University Press, 2014.

Folkerts, Jean, John Maxwell Hamilton, and Nicholas Lemann. "Educating Journalists: A New Plea for the University Tradition." Columbia University Graduate School of Journalism, October 2013. https://www.journalism.columbia.edu/system/documents/785/original/75881_JSchool_Educating_Journalists-PPG_V2-16.pdf.

Ford, Matt. "Can Bipartisanship End Mass Incarceration?" *Atlantic*, February 25, 2015. https://www.theatlantic.com/politics/archive/2015/02/can-bipartisanship-end-mass-incarceration/386012/.

Fowler, James H., and Nicholas Christakis. "Dynamic Spread of Happiness in a Large Social Network: Longitudinal Analysis of the Framingham Heart Study Social Network." *BMJ (Clinical Research Ed.)* 338, no. 7685 (January 3, 2009): 23–27. https://doi.org/10.1136/bmj.a2338.

Freeman, Linton. "A Set of Measures of Centrality Based on Betweenness." *Sociometry* 40 (1977): 35–41. http://moreno.ss.uci.edu/23.pdf.

Frey, Carl Benedikt, and Michael A. Osborne. "The Future of Employment: How Susceptible Are Jobs to Computerisation?" Oxford Martin Programme on Technology and Employment, September 17, 2013. https://www.oxfordmartin.ox.ac.uk/downloads/academic/The_Future_of_Employment.pdf.

Friedman, Sharon M., Sharon Dunwoody, and Carol L. Rogers. *Communicating Uncertainty: Media Coverage of New and Controversial Science*. New York: Routledge, 1999.

Frier, Sarah. "Facebook Reaches 2 Billion Users, Now Wants to Build Friendships." *Bloomberg News*, June 27, 2017. https://www.bloomberg.com/news/articles/2017-06-27/facebook-reaches-2-billion-users-now-wants-to-build-friendships.

Fung, Archon, Mary Graham, David Weil, and Elena Fagotto. "Transparency Policies: Two Possible Futures." *Taubman Center Policy Briefs* 1 (May 2007): 1–7. http://www.transparencypolicy.net/assets/two%20possible%20futures.pdf.

Funk, Cary, and Sara Kehaulani Goo. "A Look at What the Public Knows and Does Not Know about Science." Pew Research Center, September 10, 2015. http://www.pewinternet.org/2015/09/10/what-the-public-knows-and-does-not-know-about-science/.

Gallardo, Adrianna. "What We've Learned So Far about Maternal Mortality from You, Our Readers." ProPublica, May 18, 2017. https://www.propublica.org/article/what-weve-learned-so-far-about-maternal-mortality-from-you-our-readers.

Gallup/Knight Foundation. "American Views: Trust, Media and Democracy." 2017 Gallup/Knight Foundation Survey on Trust, Media and Democracy, January 16, 2018. https://knightfoundation.org/reports/american-views-trust-media-and-democracy.

Gallup News. "Global Warming Concern at Three-Decade High in US." March 14, 2017. http://news.gallup.com/poll/206030/global-warming-concern-three-decade-high .aspx.

Gallup News. "Media Use and Evaluation." March 14, 2017. http://news.gallup. com/poll/1663/media-use-evaluation.aspx.

Gallup News. "Satisfaction with the United States." February 2017. http://news .gallup.com/poll/1669/general-mood-country.aspx.

Galston, William A. "Political Knowledge, Political Engagement and Civic Education." *Annual Review of Political Science* 4 (June 2001): 217–234. https://doi.org/ 10.1146/annurev.polisci.4.1.217.

Gamble, James. "Wiring a Continent: The Making of the U.S. Transcontinental Telegraph." *The Californian*, 1881. http://www.telegraph-history.org/transcontinental -telegraph/.

Gamson, W. A., and A. Modigliani. "The Changing Culture of Affirmative Action." In *Research in Political Sociology*, ed. R. G. Braungart and M. M. Braungart, 137–177. Greenwich, CT: JAI Press, 1987.

Gans, Herbert J. *Democracy and the News*. Oxford: Oxford University Press, 2003.

Gans, Herbert J. "Multiperspectival News Revisited: Journalism and Representative Democracy." *Journalism* 12, no. 1 (January 2011): 3–13. https://doi.org/10.1177/ 1464884910385289.

Garimella, Kiran, Ingmar Weber, and Munmun De Choudhury. "Quote RTs on Twitter: Usage of the New Feature for Political Discourse." *Proceedings of the 8th ACM Conference on Web Science*, March 25, 2016. https://arxiv.org/pdf/1603.07933.pdf.

Gelman, Andrew. "Science Journalism and the Art of Expressing Uncertainty." The Symposium, August 4, 2013. http://www.symposium-magazine.com/science-journal ism-and-the-art-of-expressing-uncertainty/.

General Social Survey. Accessed September 1, 2017. http://gss.norc.org/.

Gentzkow, Matthew, Edward L. Glaeser, and Claudia Goldin. "The Rise of the Fourth Estate: How Newspapers Became Informative and Why It Mattered." In *Corruption and Reform: Lessons from America's Economic History*, ed. Edward L. Glaeser and Claudia Goldin, 187–230. Chicago: University of Chicago Press, 2006. http://www.nber .org/chapters/c9984.pdf.

Gentzkow, Matthew, and Jesse M. Shapiro. "What Drives Media Slant? Evidence from U.S. Daily Newspapers." *Econometrica* 78, no. 1 (January 2010): 35–71. https:// doi.org/10.3982/ECTA7195.

Gilbert, Sophie. "The Movement of #MeToo." *Atlantic*, October 16, 2017. https://www.theatlantic.com/entertainment/archive/2017/10/the-movement-of-metoo/542979/.

Gladwell, Malcolm. *The Tipping Point: How Little Things Can Make a Big Difference*. New York: Little, Brown and Company, 2000.

Glaeser, Edward L., and Cass R. Sunstein. "Why Does Balanced News Produce Unbalanced Views?" NBER Working Paper no. 18975, National Bureau of Economic Research, Cambridge, MA, April 2013. https://doi.org/10.3386/w18975.

Gleick, James. *The Information*. New York: Pantheon Books, 2011.

Goel, Sharad, Ashton Anderson, Jake Hofman, and Duncan J. Watts. "The Structural Virality of Online Diffusion." *Management Science* 62, no. 1 (July 22, 2015): 180–196. https://doi.org/10.1287/mnsc.2015.2158.

Goodman, J. David. "New York Police Twitter Strategy Has Unforeseen Consequences." *New York Times*, April 23, 2014. https://www.nytimes.com/2014/04/24/nyregion/new-york-police-twitter-strategy-has-unforeseen-consequences.html.

Goodman, J. David. "On Staten Island, Thousands Protest Police Tactics." *New York Times*, August 23, 2014. https://www.nytimes.com/2014/08/24/nyregion/on-staten-island-thousands-protest-police-tactics.html.

Google. "About." Accessed August 1, 2017. https://www.google.com/intl/en/about/.

Granovetter, Mark S. "The Strength of Weak Ties." *American Journal of Sociology* 78, no. 6 (May 1973): 1360–1380. https://doi.org/10.1086/225469.

Granovetter, Mark S. "Threshold Models of Collective Behavior." *American Journal of Sociology* 83 (1978): 1420–1443. https://doi.org/10.1086/226707.

Graves, Lucas. *Deciding What's True: The Rise of Political Fact-Checking in American Journalism*. New York: Columbia University Press, 2016.

Greenberg, Julia. "Buzzfeed Is Changing the Way It Measures Its Popularity." *Wired*, February 18, 2016. https://www.wired.com/2016/02/buzzfeed-is-changing-the-way-it-measures-its-popularity/.

Greenwood, Shannon, Andrew Perrin, and Maeve Duggan. "Social Media Update 2016." Pew Research Center, November 11, 2016. http://www.pewinternet.org/2016/11/11/social-media-update-2016/.

Grewal, David. *Network Power: The Social Dynamics of Globalization*. New Haven, CT: Yale University Press, 2008.

Grozanick, Rachel. "Pulitzer Prize Winner Dave Philipps on Using Social Media to Find Untold Stories." The GroundTruth Project, July 11, 2017. http://theground

truthproject.org/pulitzer-prize-winner-dave-philipps-on-using-social-media-to-find
-untold-stories/.

Guardian. "The Counted." *Guardian*. Accessed March 15, 2018. https://www.the
guardian.com/us-news/ng-interactive/2015/jun/01/the-counted-police-killings
-us-database.

Gunther, Albert C., and Kathleen Schmitt. "Mapping Boundaries of the Hostile
Media Effect." *Journal of Communication* 54, no. 1 (March 2004): 55–70. https://doi
.org/10.1111/j.1460-2466.2004.tb02613.x.

Gutmann, Amy, and Dennis Thompson. *Why Deliberative Democracy?* Princeton, NJ:
Princeton University Press, 2004.

Haile, Tony. "What You Think You Know about the Web Is Wrong." *Time*, March 9,
2014. http://time.com/12933/what-you-think-you-know-about-the-web-is-wrong/.

Haiman, Robert J. "Best Practices for Newspaper Journalists." Free Press/Fair Press
Project, Freedom Forum, Arlington, Virginia, September 3, 2002.

Halberstam, David. *The Powers That Be*. Chicago: First Illinois, 2000.

Halpern, Sue. "How He Used Facebook to Win." *New York Review of Books*, June 8,
2017. http://www.nybooks.com/articles/2017/06/08/how-trump-used-facebook-to
-win/.

Hannak, Aniko, Piotr Sapiezynski, Arash Molavi Kakhki, Balachander Krishnamur-
thy, David Lazer, Alan Mislove, and Christo Wilson. "Measuring Personalization of
Web Search." In *Proceedings of the 22nd International Conference on World Wide Web*,
527–538. New York: ACM, 2013. https://dl.acm.org/citation.cfm?id=2488435.

Hansen, Mark, Meritxell Roca-Sales, Jonathan M. Keegan, and George King. "Artificial Intelligence: Practice and Implications for Journalism." Tow Center for Digital
Journalism, Columbia University, September 2017. https://academiccommons
.columbia.edu/catalog/ac:gf1vhhmgs8.

Hardin, Russell. "The Crippled Epistemology of Extremism." In *Political Extremism
and Rationality*, edited by Albert Breton, Gianluigi Galeotti, Pierre Salmon, and
Ronald Wintrobe, 3–22. Cambridge: Cambridge University Press, 2002. https://doi
.org/10.1017/CBO9780511550478.002.

Hardin, Russell. *How Do You Know? The Economics of Ordinary Knowledge*. Princeton,
NJ: Princeton University Press, 2009.

Harper's New Monthly Magazine. "The Telegraph." *Harper's New Monthly Magazine*
47, no. 279 (August 1873): 332–360.

Hay, Jeremy, and Eve Pearlman. "Can People Be Civil About Polarizing Topics? 'Dialogue Journalism' Could Serve as a Roadmap." Poynter, March 1, 2018. https://www

.poynter.org/news/can-people-be-civil-about-polarizing-topics-dialogue
-journalism-could-serve-roadmap.

Hayek, F. A. "The Use of Knowledge in Society." *American Economic Review* 35, no. 4 (September 1945): 519–530. http://home.uchicago.edu/~vlima/courses/econ200/ spring01/hayek.pdf.

Hayes, Danny, and Jennifer L. Lawless. "As Local News Goes, So Goes Citizen Engagement: Media, Knowledge, and Participation in US House Elections." *Journal of Politics* 77, no. 2 (April 2015): 447–462. https://doi.org/10.1086/679749.

Head, Alison J., Michele Van Hoeck, and Kirsten Hostetler. "Why Blogs Endure: A Study of Recent College Graduates and Motivations for Blog Readership." *First Monday* 22, no. 10 (October 2, 2017). http://firstmonday.org/ojs/index.php/fm/ article/view/8065/6539.

Head, Alison J., and John Wihbey. "At Sea in a Deluge of Data." *Chronicle of Higher Education*, July 7, 2014. http://www.chronicle.com/article/At-Sea-in-a-Deluge-of-Data/ 147477/.

Head, Alison J., and John Wihbey. "The Importance of Truth Workers in an Era of Factual Recession." April 8, 2017. https://medium.com/@ajhead1/the-importance-of -truth-workers-in-an-era-of-factual-recession-7487fda8eb3b.

Head, Alison J., John Wihbey, P. Takis Metaxas, and Dan Cohen. "Examining Information Processing Practices among Young News Consumers: A National Study to Inform Journalists, Librarians, and Educators." Project Information Literacy, October 2018. http://www.projectinfolit.org/news_study.html.

Heckman, Meg, and John Wihbey. "Mobile Site Experience and Local News: Missed Innovation Opportunities?" Paper presented at the Association for Education in Journalism and Mass Communication (AEJMC), Media Management, Economics and Entrepreneurship Division, Norman, Oklahoma, March 2, 2018.

Hermida, Alfred, and Mary Lynn Young. "Finding the Data Unicorn: A Hierarchy of Hybridity in Data and Computational Journalism." *Digital Journalism* 5, no. 2 (April 7, 2016): 159–176. https://doi.org/10.1080/21670811.2016.1162663.

Hersh, Adam. "Slowing Down the Presses: The Relationship between Net Neutrality and Local News." Center for Internet and Society, Stanford University, December 12, 2017. https://cyberlaw.stanford.edu/files/publication/files/20171208-NetNeutrality andLocalNews_0.pdf.

Hill, R. A., and R. I. M. Dunbar. "Social Network Size in Humans." *Human Nature* 14, no. 1 (March 2003): 53–72. https://link.springer.com/article/10.1007/s12110-003 -1016-y.

Himelboim, Itai, Marc A. Smith, Lee Rainie, Ben Shneiderman, and Camila Espina. "Classifying Twitter Topic-Networks Using Social Network Analysis." *Social Media and Society* 3, no. 1 (February 1, 2017). https://doi.org/10.1177/2056305117691545.

Hindman, Matthew. "Journalism Ethics and Digital Audience Data." In *Remaking the News: Essays on the Future of Journalism Scholarship in the Digital Age*, ed. C. W. Anderson and Pablo J. Boczkowki, 177–194. Cambridge, MA: MIT Press, 2017.

Hochfelder, David. *The Telegraph in America, 1832–1920*. Baltimore: Johns Hopkins University Press, 2012.

Hoskins, Mike. "How Graph Analytics Deliver Deeper Understanding." *InfoWorld*, February 4, 2015. https://www.infoworld.com/article/2877489/big-data/how-graph-analytics-delivers-deeper-understanding-of-complex-data.html.

Holton, Avery E., Seth C. Lewis, and Mark Coddington. "Interacting with Audiences." *Journalism Studies* 17, no. 7 (April 2016): 849–859. https://doi.org/10.1080/1461670X.2016.1165139.

Hudson, Michael. "Panama Papers Wins Pulitzer Prize." International Consortium of Investigative Journalists, April 10, 2017. https://www.icij.org/blog/2017/04/panama-papers-wins-pulitzer-prize/.

Huey, John, Martin Nisenholtz, and Paul Sagan. "Riptide: What Really Happened to the News Business." Shorenstein Center on Media, Politics and Public Policy, September 8, 2013. https://shorensteincenter.org/2013/09/d81-riptide/.

Innis, Harold. *The Bias of Communication*. 2nd ed. Toronto: University of Toronto Press, 1999.

Jackson, Derrick Z. "Environmental Justice? Unjust Coverage of the Flint Water Crisis." Shorenstein Center on Media, Politics and Public Policy, July 2017. https://shorensteincenter.org/wp-content/uploads/2017/07/Flint-Water-Crisis-Derrick-Z-Jackson-1.pdf?x78124.

Jackson, Sarah J., and Brooke Foucault Welles. "#Ferguson Is Everywhere: Initiators in Emerging Counterpublic Networks." *Information Communication and Society* 19, no. 3 (2016): 397–418. https://doi.org/10.1080/1369118X.2015.1106571.

Jackson, Sarah J., and Brooke Foucault Welles. "Hijacking #myNYPD: Social Media Dissent and Networked Counterpublics." *Journal of Communication* 65, no. 6 (December 2015): 932–952. https://doi.org/10.1080/1369118X.2015.1106571.

James, William. *The Principles of Psychology*. Vol. 1. New York: Henry Holt & Co., 1890.

Jamieson, Kathleen Hall. "Messages, Micro-Targeting, and New Media Technologies." *Forum* 11, no. 3 (October 2013): 429–435. https://repository.upenn.edu/cgi/viewcontent.cgi?article=1363&context=asc_papers.

Janowitz, Morris. "Professional Models in Journalism: The Gatekeeper and the Advocate." *Journalism & Mass Communication Quarterly* 52, no. 4 (December 1, 1975): 618–626. https://doi.org/10.1177/107769907505200402.

Jarvis, Jeff. "Our Problem Isn't 'Fake News.' Our Problems Are Trust and Manipulation." Medium, June 12, 2017. https://medium.com/whither-news/our-problem-isnt-fake-news-our-problems-are-trust-and-manipulation-5bfbcd716440.

Jasanoff, Sheila. *Designs on Nature: Science and Democracy in Europe and the United States*. Princeton, NJ: Princeton University Press, 2005.

Jefferson, Thomas. Letter to Edward Carrington. January 16, 1787. http://tjrs.monticello.org/letter/1289.

Jefferson, Thomas. Letter to John Norvell. June 11, 1807. https://founders.archives.gov/documents/Jefferson/99-01-02-5737.

Jensen, Jakob D. "Scientific Uncertainty in News Coverage of Cancer Research: Effects of Hedging on Scientists' and Journalists' Credibility." *Human Communication Research* 34 (3) (2008): 347–369. https://doi.org/10.1111/j.1468-2958.2008.00324.x.

Jones, Alex S. *Losing the News: The Future of the News that Feeds Democracy*. Oxford: Oxford University Press, 2009.

Kahan, Dan M. "Ideology, Motivated Reasoning, and Cognitive Reflection." *Judgement and Decision Making* 8, no. 4 (July 2013): 407–424. http://journal.sjdm.org/13/13313/jdm13313.pdf.

Karpf, David. *Analytic Activism: Digital Listening and the New Political Strategy*. Oxford: Oxford University Press, 2016.

Kennedy, Dan. "Beyond the Tea Party: How the IRS Is Killing Nonprofit Media." HuffPost, July 16, 2013. https://www.huffingtonpost.com/dan-kennedy/beyond-the-tea-party-how_b_3286604.html.

Kennedy, Dan. "Tracing the Links between Civic Engagement and the Revival of Local Journalism." Nieman Lab, June 4, 2013. http://www.niemanlab.org/2013/06/tracing-the-links-between-civic-engagement-and-the-revival-of-local-journalism/.

Kennedy, Dan. *The Wired City: Reimagining Journalism and Civic Life in the Post-newspaper Age*. Amherst: University of Massachusetts Press, 2013.

Khazan, Olga. "Should Journalism Schools Require Reporters to 'Learn Code'? No." *Atlantic*, October 21, 2013. https://www.theatlantic.com/education/archive/2013/10/should-journalism-schools-require-reporters-to-learn-code-no/280711/.

Kille, Leighton Walter. "Research Chat: Social Scientist Dietram A. Scheufele on Framing and Political Communication." Journalist's Resource, Shorenstein Center on Media, Politics and Public Policy. Updated November 19, 2013. https://

journalistsresource.org/tip-sheets/research/research-chat-dietram-scheufele-framing
-political-communication.

King, Gary, Benjamin Schneer, and Ariel White. "How the News Media Activate Public Expression and Influence National Agendas." *Science* 358, no. 6364 (November 10, 2017): 776–780.

Kirchner, Lauren. "When Discrimination Is Baked into Algorithms." *Atlantic*, September 6, 2015. https://www.theatlantic.com/business/archive/2015/09/discrimination
-algorithmsdisparate-impact/403969/.

Kirkpatrick, David. *The Facebook Effect: The Inside Story of the Company that Is Connecting the World*. New York: Simon & Schuster Paperbacks, 2011.

Kirkpatrick, Keith. "Putting the Data Science into Journalism." *Communications of the ACM* 58, no. 5 (May 2015): 15–17. https://doi.org/10.1145/2742484.

Kleinberg, Jon. "Information Flow and Graph Structure in Online Social Networks." Lecture at the Becker Friedman Institute at the University of Chicago, September 24, 2016. YouTube video. https://youtu.be/iccgU1ul13E.

Klinenberg, Eric. *Fighting for Air: The Battle to Control America's Media*. New York: Metropolitan Books, 2007.

Kohut, Andrew, Robert C. Toth, and Carol Bowman. "Public Tunes Out Recent News." Times Mirror Center for the People & the Press, May 19, 1994.

The Knight Commission on the Information Needs of Communities in a Democracy. *Informing Communities: Sustaining Democracy in the Digital Age*. Washington, DC: The Aspen Institute, 2009. https://assets.aspeninstitute.org/content/uploads/files/content/docs/pubs/Informing_Communities_Sustaining_Democracy_in_the
_Digital_Age.pdf.

Kovach, Bill, and Tom Rosenstiel. *Blur: How to Know What's True in the Age of Information Overload*. New York: Bloomberg, 2011.

Kovach, Bill, and Tom Rosenstiel. *The Elements of Journalism: What Newspeople Should Know and the Public Should Expect*. New York: Three Rivers Press, 2014.

Kramer, Adam D. I., Jamie E. Guillory, and Jeffrey T. Hancock. "Experimental Evidence of Massive-Scale Emotional Contagion through Social Networks." *Proceedings of the National Academy of Sciences of the United States of America* 111, no. 24 (June 2014): 8788–8790. https://doi.org/10.1073/pnas.1320040111.

Kreiss, Daniel. "Beyond Administrative Journalism: Civic Skepticism and the Crisis in Journalism." In *The Crisis of Journalism Reconsidered: Democratic Culture, Professional Codes, Digital Future*, edited by Jeffrey C. Alexander, Elizabeth Butler Breese, and Maria Luengo, 59–76. Cambridge: Cambridge University Press, 2016.

Kreiss, Daniel. "The Media Are about Identity, Not Information." In *Trump and the Media*, edited by Pablo J. Boczkowski and Zizi Papacharissi, 93–100. Cambridge, MA: MIT Press, 2018.

Kreiss, Daniel. *Prototype Politics: Technology-Intensive Campaigning and the Data of Democracy*. Oxford: Oxford University Press, 2016.

Kuran, Timur. "Availability Cascades and Risk Regulation." University of Chicago Public Law & Legal Theory Working Paper no. 181, 2007.

Ladd, Jonathan M. *Why Americans Hate the Media and How It Matters*. Princeton, NJ: Princeton University Press, 2012.

Larson, Jeff, Surya Mattu, Lauren Kirchner, and Julia Angwin. "How We Analyzed the COMPAS Recidivism Algorithm." ProPublica, May 23, 2016. http://www .propublica.org/article/how-we-analyzed-the-compas-recidivism-algorithm/.

Lazarsfeld, Paul Felix, Bernard Berelson, and Hazel Gaudet. *The People's Choice: How the Voter Makes Up His Mind in a Presidential Campaign*. New York: Columbia University Press, 1968.

Lazarsfeld, Paul Felix, and Elihu Katz. *Personal Influence: The Part Played by People in the Flow of Mass Communications*. New York: Free Press, 1955.

Lazarsfeld, Paul, and Robert Merton. "Friendship as a Social Process: A Substantive and Methodological Analysis." In *Freedom and Control in Modern Society*, 18–66. New York: Van Nostrand, 1954.

Lazer, David. "The Rise of the Social Algorithm." *Science* 1 (May 7, 2015): 1090–1091. https://doi.org/10.1126/science.aab1422.

Lee, Angela M., Seth C. Lewis, and Matthew Powers. "Audience Clicks and News Placement: A Study of Time-Lagged Influence in Online Journalism." *Communication Research* 41, no. 4 (2014): 505–530. https://doi.org/10.1177/0093650212467031.

Lee, Eun-Ju. "That's Not the Way It Is: How User-Generated Comments on the News Affect Perceived Media Bias." *Journal of Computer-Mediated Communication* 18, no. 1 (October 10, 2012): 32–45. https://doi.org/10.1111/j.1083-6101.2012.01597.x.

Lelkes, Yphtach. "Mass Polarization: Manifestations and Measurements." *Public Opinion Quarterly* 80, no. S1 (January 1, 2016): 392–410.

Lemann, Nicholas. "The Journalistic Method: Five Principles for Blending Analysis and Narrative." Journalist's Resource, Shorenstein Center on Media, Politics and Public Policy. Updated April 8, 2016. https://journalistsresource.org/tip-sheets/ journalistic-method-tip-sheet-blending-analysis-narrative.

Lemann, Nicholas. "Solving the Problem of Fake News." *New Yorker* November 30, 2016. https://www.newyorker.com/news/news-desk/solving-the-problem-of-fake-news.

Lessig, Lawrence. "Against Transparency." *New Republic*, October 8, 2009. https://newrepublic.com/article/70097/against-transparency.

Lessig, Larry. *Code: And Other Laws of Cyberspace Version 2.0*. 2nd ed. New York: Basic Books, 2006.

LeVines, George. "How to Install and Brand Your Own Version of ChartBuilder." Storybench, January 27, 2016. http://www.storybench.org/install-brand-version-chartbuilder/.

Levitsky, Steven, and Daniel Ziblatt. *How Democracies Die: What History Reveals about Our Future*. New York: Crown, 2018.

Levy, Steven. *In the Plex: How Google Thinks, Works, and Shapes Our Lives*. New York: Simon & Schuster, 2011.

Lewis, Harry. "My REAL Contribution to the Birth of Facebook (II)." *Bits and Pieces* (blog), May 19, 2012. http://harry-lewis.blogspot.com/2012/05/my-real-contribution-to-birth-of.html.

Li, Quanzhi, Armineh Nourbakhsh, Sameena Shah, and Xiaomo Liu. "RealTime Novel Event Detection from Social Media." In *2017 IEEE 33rd International Conference on Data Engineering (ICDE)*, 1129–1139. Piscataway, NJ: IEEE, 2017. https://doi.org/10.1109/ICDE.2017.157.

Liben-Nowell, D., and J. Kleinberg. "The Link-Prediction Problem for Social Networks." *Journal of the American Society for Information Science and Technology* 58, no. 7 (May 2007): 1019–1031. https://doi.org/10.1002/asi.20591.

Licklider, J. C. R., The Council on Library Resources, and Bolt, Beranek, and Newman, Inc. *Libraries of the Future*. Cambridge, MA: MIT Press, 1965. https://books.google.com/books/about/Libraries_of_the_future.html?id=PFNpAAAAIAAJ.

Lippmann, Walter. *Liberty and the News*. New York: Harcourt, Brace and Howe, 1920. https://archive.org/details/libertynews00lippuoft.

Lippmann, Walter. *Public Opinion*. New York: Free Press Paperbacks, 1997. First published 1922.

Liu, Xiaomo, Quanzhi Li, Armineh Nourbakhsh, Rui Fang, Merine Thomas, Kajsa Anderson, Russ Kociuba, et al. "Reuters Tracer: A Large-Scale System of Detecting & Verifying RealTime News Events from Twitter." In *Proceedings of the 25th ACM International on Conference on Information and Knowledge Management*, 207–216. New York: ACM, 2017. https://doi.org/10.1145/2983323.2983363.

Liu, Xiaomo, Quanzhi Li, Armineh Nourbakhsh, and Sameena Shah. "Real-time Rumor Debunking on Twitter." In *Proceedings of the 24th ACM International Conference on Information and Knowledge Management*, 1867–1870. New York: ACM, 2015. http://dx.doi.org/10.1145/2806416.2806651.

MacDorman, Marian F., Eugene Declercq, Howard Cabral, and Christine Morton. "Recent Increases in the U.S. Maternal Mortality Rate: Disentangling Trends from Measurement Issues." *Obstetrics and Gynecology* 128, no. 3 (September 2016): 447–455. https://doi.org/10.1097/AOG.0000000000001556.

Madison, James. "Volume 5, Amendment I (Speech and Press), Document 24." *Founders' Constitution—Foundation for the United States Constitution*, January 1800. http://press-pubs.uchicago.edu/founders/documents/amendI_speechs24.html.

Madison, James. "For the National Gazette, [ca. 19 December] 1791." Founders Online, National Archives, last modified June 29, 2017. https://founders.archives .gov/documents/Madison/01-14-02-0145.

Maier, Scott R. "Accuracy Matters: A Cross-Market Assessment of Newspaper Error and Credibility." *Journalism & Mass Communication Quarterly* 82, no. 3 (2005): 533–551. https://doi.org/10.1177/107769900508200304.

Mangu-Ward, Katherine. "Wikipedia and Beyond: Jimmy Wales' Sprawling Vision." *Reason*, June 2007. http://reason.com/archives/2007/05/30/wikipedia-and-beyond.

Marconi, Francesco. *How Artificial Intelligence Will Impact Journalism*. Associated Press, 2017. https://insights.ap.org/industry-trends/report-how-artificial-intelligence -will-impact-journalism.

Marks, Joseph, Eloise Copland, Eleanor Loh, Cass R. Sunstein, and Tali Sharot. "Epistemic Spillovers: Learning Others' Political Views Reduces the Ability to Assess and Use Their Expertise in Nonpolitical Domains." Harvard Public Law Working Paper no. 18–22, April 13, 2018. https://ssrn.com/abstract=3162009.

Marks, Peter. "Word for Word/Al Gore's Senior Thesis; Even in College, the Vice President Was No Boob about the Tube." *New York Times*, July 16, 2000. https:// www.nytimes.com/2000/07/16/weekinreview/word-for-word-al-gore-s-senior-thesis -even-college-vice-president-was-no-boob.html.

Martin, Gregory J., and Ali Yurukoglu. "Bias in Cable News: Persuasion and Polarization." *American Economic Review* 107, no. 9 (September 2017): 2565–2599. https:// doi.org/10.1257/aer.20160812.

Martin, Nina, and Renee Montagne. "The Last Person You'd Expect to Die in Childbirth." ProPublica, May 12, 2017. https://www.propublica.org/article/die-in-child birth-maternal-death-rate-health-care-system.

Martin, Travis, Jake M. Hofman, Amit Sharma, Ashton Anderson, and Duncan J. Watts. "Exploring Limits to Prediction in Complex Social Systems." In *Proceedings of the 25th International Conference on World Wide Web*, 683–694. Geneva: International World Wide Web Conferences Steering Committee, 2016. https://doi.org/10.1145/ 2872427.2883001.

Mason, Lilliana. *Uncivil Agreement: How Politics Became Our Identity*. Chicago: University of Chicago Press, 2018.

McCarthy, Justin. "Trust in Mass Media Returns to All-Time Low." Gallup, September 17, 2014. http://news.gallup.com/poll/176042/trust-mass-media-returns-time -low.aspx.

McChesney, Robert W. *The Problem of the Media: U.S. Communication Politics in the 21st Century*. New York: Monthly Review Press, 2004.

McCrummen, Stephanie, Beth Reinhard, and Alice Crites. "Woman Says Roy Moore Initiated Sexual Encounter When She Was 14, He Was 32." *Washington Post*, November 9, 2017. https://www.washingtonpost.com/investigations/woman-says-roy-moore -initiated-sexual-encounter-when-she-was-14-he-was-32/2017/11/09/1f495878 -c293-11e7-afe9-4f60b5a6c4a0_story.html.

McGregor, Susan E. "A Brief History of Computer Assisted Reporting." *Tow Center for Digital Journalism* (blog), March 18, 2013. https://towcenter.org/a-brief-history-of -computer-assisted-reporting/.

McKay, Charles. *Memoirs of Extraordinary Popular Delusions and the Madness of Crowds*. London: Richard Bentley, 1841. https://www.cmi-gold-silver.com/pdf/mackaych24 51824518-8.pdf.

McPherson, Miller, Lynn Smith-Lovin, and James M. Cook. "Birds of a Feather: Homophily in Social Networks." *Annual Review of Sociology* 1 (2001), 415–444. https://doi.org/10.1146/annurev.soc.27.1.415.

Media Insight Project. "The Personal News Cycle." American Press Institute and the Associated Press-NORC Center for Public Affairs Research, March 17, 2014. https:// www.americanpressinstitute.org/publications/reports/survey-research/personal -news-cycle/.

Mele, Christopher. "Fatigued by the News? Experts Suggest How to Adjust Your Media Diet." *New York Times*, February 1, 2017. https://www.nytimes.com/2017/ 02/01/us/news-media-social-media-information-overload.html.

Mele, Nicco. *The End of Big: How the Digital Revolution Makes David the New Goliath*. New York: St. Martin's Press, 2013.

Merton, Robert K. "The Matthew Effect in Science, II: Cumulative Advantage and the Symbolism of Intellectual Property." *Isis* 79, no. 4 (December 1988): 606–623. https://doi.org/10.1086/354848.

Metaxas, Panagiotis Takis. "Web Spam, Social Propaganda and the Evolution of Search Engine Rankings." 2009. http://cs.wellesley.edu/~pmetaxas/Metaxas-Evolution SEs-LNBIP10.pdf.

Metcalfe, Bob. "Metcalfe's Law after 40 Years of Ethernet." *Computer* 46, no. 12 (2013): 26–31. https://doi.org/10.1109/MC.2013.374.

Meyer, Philip. "The Next Journalism's Objective Reporting." Nieman Reports, December 15, 2004. https://niemanreports.org/articles/the-next-journalisms-objective -reporting/.

Meyer, Philip. *Precision Journalism: A Reporter's Guide to Social Science Methods.* 4th ed. Lanham, MD: Rowan & Littlefield, 2002.

Milgram, Stanley. "The Small-World Problem." *Psychology Today* 1, no. 1 (May 1967): 61–67. https://www.cs.purdue.edu/homes/agebreme/Networks/papers/milgram 67smallworld.pdf.

Mindich, David T. Z. *Just the Facts: How "Objectivity" Came to Define American Journalism.* New York: New York University Press, 1998.

Mintz, Daniel. "3 Interesting Things Attention Minutes Have Already Taught Us." *Upworthy Insider* (blog), February 13, 2014. https://blog.upworthy.com/3-interesting -things-attention-minutes-have-already-taught-us-222f658d9b11.

Mitchell, Amy, Jeffrey Gottfried, Elisa Shearer, and Kristine Lu. "How Americans Encounter, Recall and Act upon Digital News." Pew Research Center, February 9, 2017. http://www.journalism.org/2017/02/09/how-americans-encounter-recall-and -act-upon-digital-news/.

Mitchell, Amy, Jeffrey Gottfried, Michael Barthel, and Elisa Shearer. "The Modern News Consumer." Pew Research Center, July 7, 2016. http://www.journalism.org/ 2016/07/07/trust-and-accuracy/.

Mitchell, Amy, Mark Jurkowitz, and Kenneth Olmstead. "Audience Routes: Direct, Search & Facebook." Pew Research Center, March 13, 2014. http://www.journalism .org/2014/03/13/audience-routes-direct-search-facebook/.

Munro, Geoffrey D., Peter H. Ditto, Lisa K. Lockhart, Angela Fagerlin, Mitchell Gready, and Elizabeth Peterson. "Biased Assimilation of Sociopolitical Arguments: Evaluating the 1996 U.S. Presidential Debate." *Basic and Applied Social Psychology* 24, no. 1 (2002): 15–26. https://doi.org/10.1207/S15324834BASP2401_2.

Mordecai, Adam. "9 Out of 10 Americans Are Completely Wrong About This Mind-blowing Fact." Upworthy, March 3, 2013. http://www.upworthy.com/9-out-of-10 -americans-are-completely-wrong-about-this-mind-blowing-fact-2.

Morisy, Michael. "Under Trump's First 100 Days, FOIA a Little Slower while Open Data Takes a Hit." MuckRock, April 28, 2017. https://www.muckrock.com/news/ archives/2017/apr/28/under-trumps-first-100-days-foia-little-slower-whi/.

Moses, Lucia. "As Audience Development Grows, Publishers Question Who Should Own It." Digiday, June 14, 2017. https://digiday.com/media/audience-development -grows-publishers-question/.

Moses, Lucia. "'Never More Accurate': Jack Shafer on Journalism in 2014." Digiday, December 18, 2014. https://digiday.com/media/jack-shafer-journalism-2014-never -accurate/.

Mutz, Diana C. *Hearing the Other Side: Deliberative versus Participatory Democracy*. New York: Cambridge University Press, 2006.

Nahon, Karine, and Jeff Hemsley. *Going Viral*. Cambridge: Polity Press, 2013.

National Research Council. *Network Science*. Washington, DC: National Academic Press, 2005. https://www.nap.edu/catalog/11516/network-science.

Nesmith, Susannah. "How a Regional Newspaper Pulled Off a National Investigation into Sexual Abuse by Doctors." *Columbia Journalism Review*, July 11, 2016. https:// www.cjr.org/united_states_project/atlanta_journal_constitution_doctor_sex_abuse _investigation.php.

New York Herald. "President Moves to Stop Mob Rule of Wireless." August 17, 1912.

New York Times. "Live Presidential Forecast." November 9, 2016. https://www .nytimes.com/elections/forecast/president.

New York Times. "The Times Issues Social Media Guidelines for the Newsroom." October 13, 2017. https://www.nytimes.com/2017/10/13/reader-center/social-media -guidelines.html.

Newman, M. E. J. "Modularity and Community Structure in Networks." *Proceedings of the National Academy of Sciences* 103, no. 23 (June 6, 2006): 8577–8582. https:// doi.org/10.1073/pnas.0601602103.

Nicas, Jack. "YouTube Tops 1 Billion Hours of Video a Day, on Pace to Eclipse TV." *Wall Street Journal*, February 27, 2017. https://www.wsj.com/articles/youtube-tops -1-billion-hours-of-video-a-day-on-pace-to-eclipse-tv-1488220851.

Nielsen, Rasmus Kleis. "Digital News as Forms of Knowledge: A New Chapter in the Sociology of Knowledge." In *Remaking the News: Essays on the Future of Journalism Scholarship in the Digital Age*, edited by Pablo J. Boczkowski and C. W. Anderson, 91–110. Cambridge, MA: MIT Press, 2017.

Nielsen, Rasmus Kleis, and Federica Cherubini. "Editorial Analytics: How News Media Are Developing and Using Audience Data and Metrics." Reuters Institute for the Study of Journalism, University of Oxford, 2016. https://reutersinstitute.politics. ox.ac.uk/our-research/editorial-analytics-how-news-media-are-developing-and -using-audience-data-and-metrics.

Nielsen Company. "2016 Nielsen Social Media Report." January 2017. http://www .nielsen.com/content/dam/corporate/us/en/reports-downloads/2017-reports/2016 -nielsen-social-media-report.pdf.

Nip, Joyce Y. M. "Exploring the Second Phase of Public Journalism." *Journalism Studies* 7, no. 2 (February 2007): 212–236. https://doi.org/10.1080/14616700500533528.

Nisbet, Matthew, John Wihbey, Silje Kristiansen, and Aleszu Bajak. "Funding the News: Foundations and Nonprofit Media." Shorenstein Center on Media, Politics and Public Policy, June 18, 2018. https://shorensteincenter.org/funding-the-news-foundations-and-nonprofit-media/.

Nisbet, Matthew, and Declan Fahy. "The Need for Knowledge-Based Journalism in Politicized Science Debates." *Annals of the American Academy of Political and Social Science* 658, no. 1 (February 8, 2015): 223–234. https://doi.org/10.1177/0002716214559887.

Norvell, John. Letter to Thomas Jefferson. March 5, 1809. https://founders.archives.gov/?q=Correspondent%3A%22Norvell%2C%20John%22%20Correspondent%3A%22Jefferson%2C%20Thomas%22&s=1111311111&r=3.

Nourbakhsh, Armineh, Quanzhi Li, Xiaomo Liu, and Sameena Shah. "'Breaking' Disasters: Predicting and Characterizing the Global News Value of Natural and Man-Made Disasters." Paper presented at SIGKDD, Data Science + Journalism Workshop, Halifax, Canada, August 12–17, 2017. https://www.researchgate.net/publication/318468631_Breaking_Disasters_Predicting_and_Characterizing_the_Global_News_Value_of_Natural_and_Man-made_Disasters.

NPR Press Room. "NPR Sees Large Ratings Increase." NPR Press Room, October 18, 2016.

Nyhan, Brendan. "Why the Fact-Checking at Facebook Needs to Be Checked." *New York Times*, October 23, 2017. https://www.nytimes.com/2017/10/23/upshot/why-the-fact-checking-at-facebook-needs-to-be-checked.html.

Nyhan, Brendan, and Jason Reifler. "When Corrections Fail: The Persistence of Political Misperceptions." *Political Behavior* 32, no. 2 (June 2010): 303–330. https://doi.org/10.1007/s11109-010-9112-2.

Obama, Barack. "Remarks by President Obama and Chancellor Merkel of Germany in a Joint Press Conference." Remarks given at the German Chancellory, Berlin, Germany, November 17, 2016. https://obamawhitehouse.archives.gov/the-press-office/2016/11/17/remarks-president-obama-and-chancellor-merkel-germany-joint-press.

Onnela, J. P., J. Saramaki, J. Hyvonen, G. Szabo, D. Lazer, K. Kaski, J. Kerte, and A. L. Baraba. "Structure and Tie Strengths in Mobile Communication Networks." *Proceedings of the National Academy of Sciences of the United States of America* 104, no. 18 (May 1, 2017): 7332–7336. http://www.pnas.org/content/104/18/7332.full.pdf.

Oreskes, Naomi, and Erik M. Conway. *Merchants of Doubt: How a Handful of Scientists Obscured the Truth on Issues from Tobacco Smoke to Global Warming*. New York: Bloomsbury Press, 2011.

Pachucki, Mark A., and Ronald L. Breiger. "Cultural Holes: Beyond Relationality in Social Networks and Culture." *Annual Review of Sociology* 36 (2010): 205–224. https://doi.org/10.1146/annurev.soc.012809.102615.

Pariser, Eli. *The Filter Bubble: How the New Personalized Web Is Changing What We Read and How We Think.* New York: Penguin Books Limited, 2011.

Pariser, Eli. "When the Internet Thinks It Knows You." *New York Times*, May 22, 2011. https://www.nytimes.com/2011/05/23/opinion/23pariser.html.

Park, Robert E. "News as a Form of Knowledge: A Chapter in the Sociology of Knowledge." *American Journal of Sociology* 45, no. 5 (March 1940): 669–686. https://doi.org/10.1086/218445.

Park, Robert E., Ernest W. Burgess, and Roderick D. McKenize. *The City.* Chicago: University of Chicago Press, 1967. http://www.esperdy.net/wp-content/uploads/2009/09/Park-The-City.pdf.

Patel, Sahil. "Facebook Recruits Its Top Publishers for Exclusive Shows." Digiday, May 25, 2017. https://digiday.com/media/facebook-recruits-top-video-publishers-shows/.

Patterson, Thomas E. "Doing Well and Doing Good." Shorenstein Center on Media, Politics and Public Policy, 2000. https://shorensteincenter.org/wp-content/uploads/2012/03/soft_news_and_critical_journalism_2000.pdf.

Patterson, Thomas E. *Informing the News: The Need for Knowledge-Based Journalism.* New York: Vintage Books, 2013.

Patterson, Thomas E. "News Coverage of the 2016 General Election: How the Press Failed the Voters," Shorenstein Center on Media, Politics and Public Policy, December 2016. https://shorensteincenter.org/news-coverage-2016-general-election/.

Patterson, Thomas E. *Out of Order: An Incisive and Boldly Original Critique of the News Media's Domination of America's Political Process.* New York: First Vintage Books Edition, 1994.

Patterson, Thomas E., and Philip Seib. "Informing the Public." In *The Press*, edited by Geneva Overholser and Kathleen Hall Jamieson, 189–202. Oxford: Oxford University Press, 2005.

Paul, Christopher, and Miriam Matthews. "The Russian 'Firehose of Falsehood' Propaganda Model." RAND Corporation, Perspectives, 2016. https://www.rand.org/content/dam/rand/pubs/perspectives/PE100/PE198/RAND_PE198.pdf.

Pell, M. B., and Joshua Schneyer. "The Thousands of U.S. Locales Where Lead Poisoning Is Worse than in Flint." *Reuters Investigates*, December 19, 2016. https://www.reuters.com/investigates/special-report/usa-lead-testing/.

Pentland, Alex. *Social Physics: How Good Ideas Spread—the Lessons from a New Science.* New York: Penguin Books, 2014.

Pew Research Center. "Social, Search & Direct: Pathways to Digital News." March 2014. http://assets.pewresearch.org/wp-content/uploads/sites/13/2014/03/Social-Search -and-Direct-Pathways-to-Digital-News-copy-edited.pdf.

Pew Research Center. "Trust Levels of News Sources by Ideological Group." Pew Research Center, October 20, 2014. http://www.journalism.org/2014/10/21/political -polarization-media-habits/pj_14-10-21_mediapolarization-01/.

Pew Research Center. "What the Public Knows—In Pictures, Words, Maps and Graphs." April 28, 2015.

Pew Research Center. "Younger Americans and Women Less Informed: One in Four Americans Follow National News Closely." December 28, 1995.

Pew Research Internet Project. "Why Most Facebook Users Get More than They Give." February 3, 2012. http://www.pewinternet.org/2012/02/03/why-most-face book-users-get-more-than-they-give/.

Phillips, Angela. "Transparency and the New Ethics of Journalism." *Journalism Practice* 4, no. 3 (July 8, 2010): 373–382. https://doi.org/10.1080/17512781003642972.

Picard, Robert G. "Deficient Tutelage: Challenges of Contemporary Journalism Education." Paper presented at Toward 2020: New Directions in Journalism Education, Ryerson Journalism Research Centre, Toronto, Canada, May 31, 2014. http:// ryersonjournalism.ca/2014/03/17/toward-2020-new-directions-in-journalism -education/.

Pickard, Victor. "Rediscovering the News: Journalism Studies' Three Blind Spots." In *Remaking the News: Essays on the Future of Journalism Scholarship in the Digital Age*, ed. C. W. Anderson and Pablo J. Boczkowki, 52–60. Cambridge, MA: MIT Press, 2017.

Postman, Neil. *Amusing Ourselves to Death: Public Discourse in the Age of Show Business*. New York: Viking Penguin, 1985.

Poston, Ben, and Anthony Pesce. "How We Reported This Story." *Los Angeles Times*, October 15, 2015. http://www.latimes.com/local/cityhall/la-me-crime-stats-side-201 51015-story.html.

Prell, Christina. *Social Network Analysis: History, Theory and Methodology*. Thousand Oaks, CA: SAGE Publications Limited, 2012.

Preston, Jennifer. "New York Police Twitter Backlash Spreads around the World." *New York Times*, April 23, 2014. https://thelede.blogs.nytimes.com/2014/04/23/new -york-police-twitter-backlash-spreads-around-the-world/.

Prior, Markus. *Post-Broadcast Democracy: How Media Choice Increases Inequality in Political Involvement and Polarizes Elections*. New York: Cambridge University Press, 2007.

ProPublica and the Virginian-Pilot. "Reliving Agent Orange." Series. ProPublica, 2016. https://www.propublica.org/series/reliving-agent-orange.

Putnam, Robert D. *Bowling Alone: The Collapse and Revival of American Community.* New York: Simon & Schuster, 2000.

Qiu, Xiaoyan, Diego F. M. Oliveira, Alireza Sahami Shirazi, Alessandro Flammini, and Filippo Menczer. "Limited Individual Attention and Online Virality of Low-Quality Information." *Nature* 1, no. 132 (June 26, 2017). https://www.nature.com/articles/s41562-017-0132.epdf.

Raboy, Mark. *Marconi: The Man Who Networked the World.* Oxford: Oxford University Press, 2016.

Reynolds, Ross. "KUOW Uses Speed-Dating Format to Help People Understand Each Other." *Current*, March 6, 2018. https://current.org/2018/03/kuow-uses-speed-dating-format-to-help-people-understand-each-other/.

Ricoeur, Paul. *The Course of Recognition.* Translated by David Pellauer. Cambridge, MA: Harvard University Press, 2005.

Riesman, David, with Nathan Glazer and Reuel Denney. *The Lonely Crowd: A Study of the Changing American Character.* Rev. ed. New Haven, CT: Yale University Press, 2001. First published 1961.

Robert R. McCormick Foundation. "Lessons from #10YearsUP and Resources for Audience Engagement." Robert R. McCormick Foundation. Accessed June 19, 2018. https://www.newsu.org/audience-engagement-resources.

Robinson, Sue. "Trump, Journalists, and Social Networks of Trust." In *Trump and the Media*, edited by Pablo J. Boczkowski and Zizi Papacharissi, 187–194. Cambridge, MA: MIT Press, 2018.

Rodrigues, Felippe. "How the Associated Press is Experimenting with Headlines and Modular Stories to Win Facebook." Storybench, August 26, 2017. http://www.storybench.org/how-the-associated-press-is-winning-facebook/.

Rodrigues, Felippe. "NodeXL's Marc Smith on How Mapping Virtual Crowd Networks Is Relevant to Journalism." Storybench, November 27, 2017. http://www.storybench.org/marc-smith-mapping-social-networks/.

Romano, Tricia. "In Seattle Art World, Women Run the Show." *Seattle Times*, June 18, 2016. https://www.seattletimes.com/entertainment/visual-arts/in-seattle-art-world-women-run-the-show/.

Romero, Daniel M., Brendan Meeder, and Jon Kleinberg. "Differences in the Mechanics of Information Diffusion across Topics: Idioms, Political Hashtags, and Complex Contagion on Twitter." In *Proceedings of the 20th International Conference on*

World Wide Web, 695–704. New York: ACM, 2011. https://doi.org/10.1145/1963405.1963503.

Rosen, Jay. "The People Formerly Known as the Audience." Press Think, June 27, 2006. http://archive.pressthink.org/2006/06/27/ppl_frmr.html.

Rosen, Jay. *What Are Journalists For?* New Haven, CT: Yale University Press, 1999.

Rosenberry, Jack, and Burton St. John III. "Introduction: Public Journalism Values in an Age of Media Fragmentation." In *Public Journalism 2.0: The Promise and Reality of a Citizen-Engaged Press*, edited by Jack Rosenberry and Burton St. John III, 1–8. New York: Routledge, 2010.

Rosenson, Beth A. "Against Their Apparent Self-Interest: The Authorization of Independent State Legislative Ethics Commissions." *State Politics and Policy Quarterly* 3, no. 1 (Spring 2003): 42–65.

Rosenstiel, Tom. "News as Collaborative Intelligence: Correcting the Myths about News in the Digital Age." Center for Effective Public Management at Brookings, June 30, 2015. https://www.brookings.edu/research/news-as-collaborative-intelligence-correcting-the-myths-about-news-in-the-digital-age/.

Rosenstiel, Tom. "Solving Journalism's Hidden Problem: Terrible Analytics." Center for Effective Public Management at Brookings, February 2016. https://www.brookings.edu/wp-content/uploads/2016/07/Solving-journalisms-hidden-problem.pdf.

Rosenstiel, Tom. "Why 'Be Transparent' Has Replaced 'Act Independently' as a Guiding Journalism Principle." Poynter, September 16, 2013. https://www.poynter.org/news/why-be-transparent-has-replaced-act-independently-guiding-journalism-principle.

Russell, Adrienne. *Networked: A Contemporary History of News in Transition.* Cambridge: Polity Press, 2011.

Ryan, Michael, and Les Switzer. "Balancing Arts and Sciences, Skills, and Conceptual Content." *Journalism & Mass Communication Educator* 56, no. 2 (June 1, 2001): 55–68. https://doi.org/10.1177/107769580105600205.

Sampson, Zachary, and Lisa Gartner. "Hot Wheels." *Tampa Bay Times*, April 26, 2017. http://www.tampabay.com/projects/2017/investigations/florida-pinellas-auto-theft-kids-hot-wheels/car-theft-epidemic/.

Sandvig, Christian. "The Facebook 'It's Not Our Fault' Study." *Social Media Collective Research Blog*, May 7, 2015. https://socialmediacollective.org/2015/05/07/the-facebook-its-not-our-fault-study/.

Santana, Arthur D., and Toby Hopp. "Tapping into a New Stream of (Personal) Data: Assessing Journalists' Different Use of Social Media." *Journalism & Mass Communication Quarterly* 93, no. 2 (2016): 383–408. https://doi.org/10.1177/1077699016637105.

Satija, Neena, Kiah Collier, Al Shaw, and Jeff Larson. "Hell and High Water." ProPublica, March 3, 2016. https://projects.propublica.org/houston/.

Schatz, Bruce R. "Information Retrieval in Digital Libraries: Bringing Search to the Net." *Science* 275 (January 17, 1997): 327–334. http://citeseerx.ist.psu.edu/viewdoc/download?doi=10.1.1.24.3842&rep=rep1&type=pdf.

Schmidt, Ana Lucía, Fabiana Zollo, Michela Del Vicario, Alessandro Bessi, Antonio Scala, Guido Caldarelli, H. Eugene Stanley, and Walter Quattrociocchi. "Anatomy of News Consumption on Facebook." *Proceedings of the National Academy of Sciences of the United States of America* 114, no. 12 (March 1, 2017): 3035–3039. https://doi.org/10.1073/pnas.1617052114.

Schmidt, Christine. "What Strategies Work Best for Increasing Trust in Local Newsrooms? Trusting News Has Some Ideas." Nieman Lab, February 16, 2018. http://www.niemanlab.org/2018/02/what-strategies-work-best-for-increasing-trust-in-local-newsrooms-trusting-news-has-some-ideas/.

Schmitt, Frederick F., ed. *Socializing Epistemology: The Social Dimensions of Knowledge.* London: Rowman & Littlefield, 1994.

Schroeder, Christopher M. "BuzzFeed Wins the Internet Daily: Here's What Its Boss Thinks Is Next." Recode, December 19, 2016. https://www.recode.net/2016/12/19/14010044/buzzfeed-wins-internet-future-of-media-online-social.

Schudson, Michael. "The 'Lippmann-Dewey Debate' and the Invention of Walter Lippmann as an Anti-Democrat 1986–1996." *International Journal of Communication* 2 (September 2008): 1032–1042.

Schudson, Michael. *The Power of News.* Cambridge, MA: Harvard University Press, 1982.

Schudson, Michael. *The Rise of the Right to Know: Politics and the Culture of Transparency, 1945–1975.* Cambridge, MA: Harvard University Press, 2015.

Schudson, Michael. "Question Authority: A History of the News Interview in American Journalism, 1860s–1930s." *Media, Culture & Society* 16 (1994): 565–587.

Schudson, Michael. *The Sociology of News.* New York: W. W. Norton & Company, 2003.

Shaker, Lee. "Dead Newspapers and Citizens' Civic Engagement." *Political Communication* 31, no. 1 (2014): 131–148. https://doi.org/10.1080/10584609.2012.762817.

Sharkey, Patrick, Gerard Torrats-Espinosa, and Delaram Takyar. "Community and the Crime Decline: The Causal Effect of Local Nonprofits on Violent Crime." *American Sociological Review* 82, no. 6 (October 25, 2017): 1214–1240. https://doi.org/10.1177/0003122417736289.

Shaw, Al, and Jeff Larson. "How We Made Hell and High Water." ProPublica, March 3, 2016. https://www.propublica.org/nerds/how-we-made-hell-and-high-water.

Shearer, Elisa, and Jeffery Gottfried. "News Use across Social Media Platforms 2017." Pew Research Center, September 7, 2017. http://www.journalism.org/2017/09/07/news-use-across-social-media-platforms-2017/.

Shifman, Limor. *Memes in Digital Culture*. Cambridge, MA: MIT Press, 2014.

Shirky, Clay. *Here Comes Everybody: The Power of Organizing without Organizations*. New York: Penguin, 2008.

Shoemaker, Pamela J., Martin Eichholz, Eunyi Kim Director, and Brenda Wrigley. "Individual and Routine Forces in Gatekeeping." *Journalism & Mass Communication Quarterly* 78, no. 2 (June 2001): 233–246. https://doi.org/10.1177/107769900107800202.

Short, James E. *How Much Media? 2015 Report on American Consumers*. Los Angeles: Institute for Communication Technology Management, University of Southern California, 2013.

Schulhofer-Wohl, Sam, and Miguel Garrido. "Do Newspapers Matter? Evidence from the Closure of *The Cincinnati Post*." Princeton University, Woodrow Wilson School of Public and International Affairs, Discussion Papers in Economics, No. 236, October 2009.

Silver, Nate. "The Media Has a Probability Problem." FiveThirtyEight, September 21, 2017. https://fivethirtyeight.com/features/the-media-has-a-probability-problem/.

Simo, Fidji. "Introducing: The Facebook Journalism Project." *Facebook Journalism Project* (blog), January 11, 2017. https://media.fb.com/2017/01/11/facebook-journalism-project/.

Sims, Janell. "BuzzFeed: The New Newsroom ... Is It the Future?" Shorenstein Center on Media, Politics and Public Policy, February 25, 2014. https://shorensteincenter.org/ben-smith/.

Skeem, Jennifer L., and Christopher T. Lowenkamp. "Risk, Race, & Recidivism: Predictive Bias and Disparate Impact." Social Science Research Network, June 15, 2016. https://doi.org/10.2139/ssrn.2687339.

Skocpol, Theda. *Diminished Democracy: From Membership to Management in American Civic Life*. Norman: University of Oklahoma Press, 2003.

Sloman, Steven, and Philip Fernbach. *The Knowledge Illusion: Why We Never Think Alone*. New York: Riverhead Books, 2017.

Small, Mario Luis. "Weak Ties and the Core Discussion Network: Why People Regularly Discuss Important Matters with Unimportant Alters." *Social Networks* 35 (2013): 470–483. https://doi.org/10.1016/j.socnet.2013.05.004.

Smith, Adam. "A Close Friend, or a Critical One?" Medium, October 25, 2017. https://medium.com/severe-contest/a-close-friend-or-a-critical-one-f4d1cc058114.

Smith, Ben, and Byron Tau. "Birtherism: Where It All Began." *Politico*, April 22, 2011. https://www.politico.com/story/2011/04/birtherism-where-it-all-began-053563.

Smith, Marc A., Lee Rainie, Ben Shneiderman, and Itai Himelboim. "Mapping Twitter Topic Networks: From Polarized Crowds to Community Clusters." Pew Research Center, February 20, 2014. http://www.pewinternet.org/2014/02/20/mapping-twitter-topic-networks-from-polarized-crowds-to-community-clusters/.

Spiegelhalter, David, Mike Pearson, and Ian Short. "Visualizing Uncertainty about the Future." *Science* 333, no. 6048 (September 9, 2011): 1393–1400. http://science.sciencemag.org/content/333/6048/1393.full.

Spinner, Jackie. "The Big Conundrum: Should Journalists Learn Code?" *American Journalism Review*, September 24, 2014. http://ajr.org/2014/09/24/should-journalists-learn-code/.

Stanford InfoLab. "The PageRank Citation Ranking: Bringing Order to the Web." Technical report. Stanford InfoLab, 1998. http://ilpubs.stanford.edu:8090/422/1/1999-66.pdf.

Starbird, Kate. "Examining the Alternative Media Ecosystem through the Production of Alternative Narratives of Mass Shooting Events on Twitter." Association for the Advancement of Artificial Intelligence, 2007. http://faculty.washington.edu/kstarbi/Alt_Narratives_ICWSM17-CameraReady.pdf.

Starr, Paul. *The Creation of the Media: Political Origins of Modern Communications*. New York: Basic Books, 2004.

Steffens, Lincoln. *The Autobiography of Lincoln Steffens*. New York: Harcourt, Brace, and World, 1936.

Stephens, Mitchell. *Beyond News: The Future of Journalism*. New York: Columbia University Press, 2014.

Storybench. "Reinventing Local TV News Project." Storybench, Northeastern University School of Journalism. Accessed June 19, 2018. http://www.storybench.org/category/tvnews/.

Strauss, Daniel. "Moore Lashes Out at Washington Post." Politico, November 11, 2017. https://www.politico.com/story/2017/11/11/roy-moore-washington-post-244805.

Stray, Jonathan. "The Age of the Cyborg." *Columbia Journalism Review*, Fall/Winter 2016. https://www.cjr.org/analysis/cyborg_virtual_reality_reuters_tracer.php.

Stray, Jonathan. "Network Analysis in Journalism: Practices and Possibilities." Paper presented at Data Science + Journalism Workshop, ACM SIGKDD in Halifax, Canada,

August 2017. https://drive.google.com/file/d/0B8CcT_0LwJ8QMzFjTWxLSFVkVTg/view.

Streitfeld, David. "'The Internet is Broken': @ev Is Trying to Salvage It." *New York Times*, May 20, 2017. https://www.nytimes.com/2017/05/20/technology/evan-williams-medium-twitter-internet.html.

Stroud, Natalie Jomini. "Polarization and Partisan Selective Exposure." *Journal of Communication* 60 (2010): 556–576. https://doi.org/10.1111/j.1460-2466.2010.01497.x.

Stroud, Natalie Jomini. *Niche News: The Politics of News Choice*. New York: Oxford University Press, 2011.

Sullivan, Danny. "Google Now Handles at Least 2 Trillion Searches per Year." Search Engine Land, May 24, 2016. https://searchengineland.com/google-now-handles-2-999-trillion-searches-per-year-250247.

Sunstein, Cass R. *#Republic: Divided Democracy in the Age of Social Media*. Princeton, NJ: Princeton University Press, 2017.

Sunstein, Cass R. *Going to Extremes: How Like Minds Unite and Divide*. New York: Oxford University Press, 2009.

Sunstein, Cass R. *Republic.com*. Princeton, NJ: Princeton University Press, 2001.

Sunstein, Cass R., Sebastian Bobadilla-Suarez, Stephanie C. Lazzaro, and Tali Sharot. "How People Update Beliefs about Climate Change: Good News and Bad News." Social Science Research Network, September 2, 2016. https://ssrn.com/abstract=2821919.

Sunstein, Cass R., and Adrian Vermeule. "Conspiracy Theories." John M. Olin Program in Law and Economics Working Paper no. 387, University of Chicago, Chicago, 2008.

Tak, Susanne, Alexander Toet, and Jan van Erp. "The Perception of Visual Uncertainty Representation by Non-experts." *IEEE Transactions on Visualization and Computer Graphics* 20, no. 6 (June 2014): 935–943. https://ieeexplore.ieee.org/abstract/document/6654171/.

Tandoc, Edson C., Jr., and Joy Jenkins. "The Buzzfeedication of Journalism? How Traditional News Organizations Are Talking about a New Entrant to the Journalistic Field Will Surprise You!," *Journalism* 18, no. 4 (December 24, 2015). http://journals.sagepub.com/doi/abs/10.1177/1464884915620269.

Tandoc, Edson C., Jr., and Tim P. Vos. "The Journalist Is Marketing the News: Social Media in the Gatekeeping Process." *Journalism Practice* 10, no. 8 (2016): 950–966. https://doi.org/10.1080/17512786.2015.1087811.

Taylor, Charles. "The Politics of Recognition." In *Multiculturalism: Examining the Politics of Recognition*, edited by A. Gutmann, 25–73. Princeton, NJ: Princeton University Press, 1992.

Tewksbury, David, and Julius Matthew Riles. "Polarization as a Function of Citizen Predispositions and Exposure to News on the Internet." *Journal of Broadcasting & Electronic Media* 59, no. 3 (2015): 381–398. https://doi.org/10.1080/08838151.2015.1054996.

Thomson Reuters. "The Making of Reuters News Tracer." Thomson Reuters, April 25, 2017. https://blogs.thomsonreuters.com/answerson/making-reuters-news-tracer/.

Thompson, Clive. "Is the Tipping Point Toast?" Fast Company, February 1, 2008. https://www.fastcompany.com/641124/tipping-point-toast.

Tofel, Richard J. "Non-profit Journalism: Issues around Impact." ProPublica. Accessed May 1, 2018. https://s3.amazonaws.com/propublica/assets/about/LFA_ProPublica-white-paper_2.1.pdf.

Tofel, Richard J. "Why Corruption Grows in Our States: Fewer Reporters and Remote State Capitals." *Daily Beast*, May 24, 2010. https://www.thedailybeast.com/why-corruption-grows-in-our-states-fewer-reporters-and-remote-state-capitals.

Tufekçi, Zeynep. "After the Protests." *New York Times*, March 14, 2009. https://www.nytimes.com/2014/03/20/opinion/after-the-protests.html.

Tufekci, Zeynep. "How Facebook's Algorithm Suppresses Content Diversity (Modestly) and How the Newsfeed Rules Your Clicks." *The Message* (blog), May 7, 2015. https://medium.com/message/how-facebook-s-algorithm-suppresses-content-diversity-modestly-how-the-newsfeed-rules-the-clicks-b5f8a4bb7bab.

Tufekçi, Zeynep. *Twitter and Tear Gas: The Power and Fragility of Networked Protest.* New Haven, CT: Yale University Press, 2017.

Tully, Melissa, Shawn Harmsen, Jane B. Singer, and Brian Ekdale. "Case Study Shows Disconnect on Civic Journalism's Role." *Newspaper Research Journal* 38, no. 4 (Fall 2017): 484–496. https://doi.org/10.1177/0739532917739881.

Twenge, Jean M., Keith W. Campbell, and Nathan T. Carter. "Declines in Trust in Others and Confidence in Institutions among American Adults and Late Adolescents, 1972–2012." *Psychological Science* 25, no. 10 (2014): 1914–1923. https://doi.org/10.1177/0956797614545133.

TwitterTrails. "Claim: Melania Trump Had an Exorcist in the White House." TwitterTrails, February 22, 2018. http://twittertrails.wellesley.edu/~trails/stories/investigate.php?id=6911497280.

Underwood, Mimi. "Updating Our Search Quality Rating Guidelines." *Google Webmaster Central Blog*, November 19, 2015. https://webmasters.googleblog.com/2015/11/updating-our-search-quality-rating.html.

United States Kerner Commission. *Report of the National Advisory Commission on Civil Disorders*. New York: Bantam, 1968.

Uscinski, Joseph E. *The People's News: Media, Politics, and the Demands of Capitalism.* New York: New York University Press, 2014.

Vaidhyanathan, Siva. *The Googlization of Everything (And Why We Should Worry).* Berkeley: University of California Press, 2011.

Van der Haak, Bregtje, Michael Parks, and Manuel Castells. "The Future of Journalism: Networked Journalism." *International Journal of Communication* 6 (2012): 1–20. http://ijoc.org/index.php/ijoc/article/view/1750/832.

Vosoughi, Soroush, Deb Roy, and Sinan Aral. "The Spread of True and False News Online." *Science* 359, no. 6380 (March 2018): 1146–1151. https://doi.org/10.1126/science.aap9559.

Wagner, Kurt. "Facebook Found a New Way to Identify Spam and False News Articles in Your News Feed." Recode, June 30, 2017. https://www.recode.net/2017/6/30/15896544/facebook-fake-news-feed-algorithm-update-spam.

Waldman, Steven, and the Working Group on Information Needs of Communities Federal Communications Commission. *Information Needs of Communities: The Changing Media Landscape in a Broadband Age*. Durham, NC: Carolina Academic Press, 2011.

Waldrop, M. Mitchell. "The Chips Are Down for Moore's Law." *Nature* 530 (February 2016): 144–147.

Wallace, David Foster. "Kenyon Commencement Address." Lecture, May 21, 2005. https://web.ics.purdue.edu/~drkelly/DFWKenyonAddress2005.pdf.

Wardle, Claire. "6 Types of Misinformation Circulated This Election Season." *Columbia Journalism Review*, November 18, 2016. https://www.cjr.org/tow_center/6_types_election_fake_news.php?CJR.

Washington Post. "Fatal Force." Database. *Washington Post*. Accessed June 19, 2018. https://www.washingtonpost.com/policeshootings/.

Watts, Duncan J. *Everything Is Obvious: How Common Sense Fails Us*. New York: Crown Business, 2011.

Watts, Duncan J. *Six Degrees: The Science of a Connected Age*. New York: W. W. Norton & Co., 2003.

Watts, Duncan J., and Steven H. Strogatz. "Collective Dynamics of 'Small-World' Networks." *Nature* 393 (June 4, 1998): 440–442. https://doi.org/10.1038/30918.

Weaver, David H., and Lars Willnat. "Changes in US Journalism: How Do Journalists Think about Social Media?" *Journalism Practice* 10, no. 7 (2016): 844–855. https://doi.org/10.1080/17512786.2016.1171162.

Webb, Amy. "2018 Tech Trends for Journalism and Media." Future Today Institute, October 2017. https://futuretodayinstitute.com/2018-tech-trends-for-journalism-and-media/.

Webster, James G. *The Marketplace of Attention: How Audiences Take Shape in a Digital Age*. Cambridge, MA: MIT Press, 2014.

Weedon, Jen, William Nuland, and Alex Stamos. "Information Operations and Facebook." Facebook, April 27, 2017. https://fbnewsroomus.files.wordpress.com/2017/04/facebook-and-information-operations-v1.pdf.

Weinberger, David. *Small Pieces Loosely Joined: A Unified Theory of the Web*. New York: Basic Books, 2002.

Weinberger, David. *Too Big to Know: Rethinking Knowledge Now that Facts Aren't the Facts, Experts Are Everywhere, and the Smartest Person in the Room Is the Room*. New York: Basic Books, 2012.

Weiss, Charles. "Expressing Scientific Uncertainty." *Law, Probability and Risk* 2, no. 1 (March 2003): 25–46.

Wellman, Barry, Anabel Quan-Haase, Jeffrey Boase, Wenhong Chen, Keith Hampton, Isabel Diaz, and Kakuko Miyata. "The Social Affordances of the Internet for Networked Individualism." *Journal of Computer-Mediated Communication* 8, no. 3 (April 2003). https://doi.org/10.1111/j.1083-6101.2003.tb00216.x.

Wellman, Barry, and Lee Rainie. *Networked: The New Social Operating System*. Cambridge, MA: MIT Press, 2012.

Weng, Lilian, Filippo Menczer, and Yong-Yeol Ahn. "Virality Prediction and Community Structure in Social Networks." *Scientific Reports* 3 (2013): 2522. https://doi.org/10.1038/srep02522.

Wihbey, John. "The Challenges of Democratizing News and Information: Examining Data on Social Media, Viral Patterns and Digital Influence." Shorenstein Center on Media, Politics and Public Policy Discussion Paper Series D-85, June 6, 2014. https://shorensteincenter.org/d85-wihbey/.

Wihbey, John. "Does the Secret to Social Networking Lie in the Remote Jungle?" *Boston Globe*, October 4, 2015. https://www.bostonglobe.com/ideas/2015/10/03/does-secret-social-networking-lie-remote-jungle/RnSfIEqW67ZiNHhOmwXIvJ/story.html.

Wihbey, John. "Guide to Critical Thinking, Research, Data and Theory: Overview for Journalists." Journalist's Resource, Shorenstein Center on Media, Politics and Public Policy. Updated March 6, 2015. https://journalistsresource.org/tip-sheets/research/guide-academic-methods-critical-thinking-theory-overview-journalists.

Wihbey, John. "Journalists' Use of Knowledge in an Online World: Examining Reporting Habits, Sourcing Practices and Institutional Norms." *Journalism Practice* 11, no. 10 (November 3, 2016): 1267–1282. https://doi.org/10.1080/17512786.2016.1249004.

Wihbey, John. "Making Federal Data More Useful and Accessible to Fuel Media and Democracy: A Report for the Federal Committee on Statistical Methodology." December 12, 2014. https://ssrn.com/abstract=2567932.

Wihbey, John. "Research Chat: Nicholas Lemann on Journalism, Scholarship, and More Informed Reporting." Journalist's Resource, Shorenstein Center on Media, Politics and Public Policy. Updated December 7, 2011. https://journalistsresource.org/tip-sheets/research/nicholas-lemann-journalism-scholarship-reporting.

Wihbey, John, and Michael Beaudet. "State-Level Policies for Personal Financial Disclosure: Exploring the Potential for Public Engagement on Conflict-of-Interest Issues." Paper presented at the Association for Education in Journalism and Mass Communication, Law and Policy Division, Chicago, Illinois, August 2017.

Wihbey, John, and Michael Beaudet. "Why It's So Hard to See Politicians' Financial Data." *New York Times*, October 4, 2016. https://www.nytimes.com/2016/10/04/opinion/why-its-so-hard-to-see-politicians-financial-data.html.

Wihbey, John, Michael Beaudet, and Pedro Miguel Cruz. "There Are Huge Holes in How the U.S. States Investigate Politicians' Conflicts of Interest." *Washington Post*, January 12, 2017. https://www.washingtonpost.com/news/monkey-cage/wp/2017/01/12/how-do-states-investigate-officials-potential-conflicts-of-interest-we-checked/.

Wihbey, John, and Mark Coddington. "Knowing the Numbers: Assessing Attitudes among Journalists and Educators about Using and Interpreting Data, Statistics, and Research." *#ISOJ Journal* 7, no. 1 (April 2017). http://isoj.org/research/knowing-the-numbers-assessing-attitudes-among-journalists-and-educators-about-using-and-interpreting-data-statistics-and-research/.

Wihbey, John, Kenny Joseph, Thalita Coleman, and David Lazer. "Exploring the Ideological Nature of Journalists' Social Networks on Twitter and Associations with News Story Content." Paper presented at Data Science + Journalism Workshop, ACM SIGKDD in Halifax, Canada, August 2017. http://lazerlab.net/publication/exploring-ideological-nature-journalists-social-networks-twitter-and-associations-news.

Willens, Max. "How BuzzFeed Gets Its Employees Data-Focused." DigiDay, March 27, 2017. https://digiday.com/media/buzzfeed-gets-employees-data-focused/.

Willnat, Lars, and David H. Weaver. "The American Journalist in the Digital Age: Key Findings." School of Journalism, Indiana University, 2014. http://archive.news .indiana.edu/releases/iu/2014/05/2013-american-journalist-key-findings.pdf.

Wittgenstein, Ludwig. *Philosophical Investigations*. 3rd ed. Translated by G. E. M. Anscombe. London: Pearson, 1973. First published 1953.

Wu, Tim. *The Attention Merchants: The Epic Scramble to Get Inside Our Heads*. New York: Borzoi Books, 2016.

Wu, Tim. "Is the First Amendment Obsolete?" Knight First Amendment Institute, Emerging Threats, September 2017. https://knightcolumbia.org/content/tim-wu -first-amendment-obsolete.

Wu, Tim. *The Master Switch: The Rise and Fall of Information Empires*. New York: Vintage Books, 2011.

Yettick, Holly. "One Small Droplet: News Media Coverage of Peer-Reviewed and University-Based Education Research and Academic Expertise." *Education Researcher* 44, no. 3 (2015): 173–184. https://doi.org/10.3102/0013189X15574903.

Zachary, G. Pascal. "To Prepare 21st-Century Journalists, Help Students Become Experts." *Chronicle of Higher Education*, December 1, 2014. http://www.chronicle .com/article/To-Prepare-21st-Century/150267/.

Zamith, Rodrigo. "Quantified Audiences in News Production." *Digital Journalism* 6, no. 4 (2018): 418–435. https://doi.org/10.1080/21670811.2018.1444999.

Zittrain, Jonathan. "Facebook Could Decide an Election without Anyone Ever Finding Out." *New Republic*, June 1, 2014. https://newrepublic.com/article/117878/ information-fiduciary-solution-facebook-digital-gerrymandering.

Zittrain, Jonathan. "The Generative Internet." *Harvard Law Review* 119, no. 7 (2006): 1974–2040. https://doi.org/10.1145/1435417.1435426.

Zuckerberg, Mark. "Building Global Community." Facebook, February 16, 2017. https://www.facebook.com/notes/mark-zuckerberg/building-global-community/ 10154544292806634/.

Zuckerberg, Mark. Lecture at CS50, December 7, 2005. YouTube video. https://www .youtube.com/watch?v=xFFs9UgOAlE&t=358s.

Zuckerman, Ethan. *Digital Cosmopolitans in the Age of Connection*. New York: W. W. Norton & Co., 2013.

Zuckerman, Ethan. "New Media, New Civics?" *Policy and Internet* 6, no. 2 (June 2014): 151–168. https://doi.org/10.1002/1944-2866.POI360.

Index